I0056013

Crop Protection and Management

Crop Protection and Management

Sarah Lancaster

SYRAWOOD
PUBLISHING HOUSE

New York

Published by Syrawood Publishing House,
750 Third Avenue, 9th Floor,
New York, NY 10017, USA
www.syrawoodpublishinghouse.com

Crop Protection and Management
Sarah Lancaster

© 2022 Syrawood Publishing House

International Standard Book Number: 978-1-64740-060-6 (Hardback)

This book contains information obtained from authentic and highly regarded sources. All chapters are published with permission under the Creative Commons Attribution Share Alike License or equivalent. A wide variety of references are listed. Permissions and sources are indicated; for detailed attributions, please refer to the permissions page. Reasonable efforts have been made to publish reliable data and information, but the authors, editors and publisher cannot assume any responsibility for the validity of all materials or the consequences of their use.

Trademark Notice: Registered trademark of products or corporate names are used only for explanation and identification without intent to infringe.

Cataloging-in-Publication Data

Crop protection and management / Sarah Lancaster.
    p. cm.
Includes bibliographical references and index.
ISBN 978-1-64740-060-6
1. Crops. 2. Plants, Protection of. 3. Cropping systems--Management.
4. Agricultural systems--Management. I. Lancaster, Sarah.
S602.5 .C76 2022
630--dc23

Table of Contents

Permissions

Index

Preface

It is with great pleasure that I present this book. It has been carefully written after numerous discussions with my peers and other practitioners of the field. I would like to take this opportunity to thank my family and friends who have been extremely supporting at every step in my life.

The science and practice of protecting crops from diseases, weeds and other vertebrate and invertebrate pests, is known as crop protection. There are various methods and approaches to plant protection. Pesticides are also widely used for crop protection. It includes insecticides, herbicides and fungicides. Biological pest control is a method of protecting pests by using other organisms. These include cover crops, beetle banks and trap crops. Barrier-based approaches such as bird netting and agrotextiles are also used. Several other techniques include bird scarers, genetic modification and plant breeding. Crop management aims to improve the growth, development as well as yield of crops. Crop management practices vary based on the type of crops. The biological characteristics of the crops are the main factors behind timing, combination and sequence of the practices used in its management. This book is a compilation of chapters that discuss the most vital concepts in the field of crop protection and management. The topics included herein are of utmost significance and bound to provide incredible insights to readers. This book is a complete source of knowledge on the present status of this important field.

The chapters below are organized to facilitate a comprehensive understanding of the subject:

Chapter – Crops and Crop Management

The plant or animal product that is extensively grown and harvested for profit is known as a crop. Cereals crop, millet crop, perennial crop, cash crop, forage crop, row crop are a few types of crops. The practices used to improve the development, growth and yield of agricultural crops fall under crop management. This is an introductory chapter which will briefly introduce all the significant aspects of crop and crop management.

Chapter – Types of Cropping

Multicropping, intercropping and monocropping are the main types of cropping. Multicropping includes the production of more than one crop in the same piece of land in the same seasons and intercropping is the practice of growing different kinds of crop in close proximity. The topics elaborated in this chapter will help in gaining a better perspective about these various types of cropping.

Chapter – Crop Diseases

Crop diseases are often caused by pathogens such as fungi and virus. Gibberella fujikuroi, Athelia rolfsii, false smut, sheath rot, helminthosporium oryzae, bacterial leaf blight are the most common diseases that affect crops. This chapter has been carefully written to provide an easy understanding of these different types of crop diseases.

Chapter – Crop Protection

The general practice that is used to protect crop yields from harmful agents like weeds, pests and plant diseases is referred to as crop protection. It includes pesticides, biopesticides, herbicides, insecticides, neonicotinoid and fungicides which are used to protect crops. This chapter discusses in detail about these agents of crop protection.

Chapter – Weed Control

Weed control is a method of pest control that attempts to stop noxious weeds from competing for resources with desired flora and fauna. Various physical and mechanical methods are used to control weeds such as coverings, manual removal, thermal, tillage and seed targeting. This chapter closely examines these key concepts of weed control to provide an extensive understanding of the subject.

Chapter – Pest Control

The organisms that reduce the quality, availability and value of human resources are classified as pests. Insects, mites, rodents, molluscs and birds are the various types of pests. The following chapter elucidates the different techniques and methods associated with pest control such as physical pest control, fumigation, biological pest control, etc.

Sarah Lancaster

1
Crops and Crop Management

The plant or animal product that is extensively grown and harvested for profit is known as a crop. Cereals crop, millet crop, perennial crop, cash crop, forage crop, row crop are a few types of crops. The practices used to improve the development, growth and yield of agricultural crops fall under crop management. This is an introductory chapter which will briefly introduce all the significant aspects of crop and crop management.

Crops

Crop is a plant or plant product that can be grown and harvested extensively for profit or subsistence. By use, crops fall into six categories: food crops, for human consumption; feed crops, for livestock consumption; fibre crops, for cordage and textiles; oil crops, for consumption or industrial uses; ornamental crops, for landscape gardening; and industrial and secondary crops, for various personal and industrial uses.

Classification of Crops

There are seven categories for classification of the crops:

- Classification Based on Season:
    - Kharif Crops : The Kharif/monsoon crop is sown with the beginning of the first rains in June to July, during the south-west monsoon season and harvest in the month of October. The term Kharif means "autumn" in Arabic, e.g. Paddy, Maize, Soybean, Sorghum, Groundnut, Cotton, Urd etc.
    - Rabi Crops : The Rabi/winter crop or spring harvest crop is sown in the month of November during the north-east monsoon season and harvest in the month of February. The term Rabi means "spring" in Arabic, e.g. Wheat, Chickpea, Pea, Sunflower, Safflower, Mustard etc.

- Zaid Crops : The Zaid/summer crop is mainly sown in the month of March and harvested in May. Requires warm day weather for major growth period and longer day length for flowering, e.g. Paddy, Groundnut, Moong, Sunflower, Maize, Watermelon, Pumpkins and Gourds.

- Classification According to Taxonomy:

  - Asteraceae/Composite : Sunflower, Safflower, Niger,

  - Cruciferae : Mustard, Radish, Cabbage, Cauliflower,

  - Cucurbitaceous : Bottle gourd, Bitter ground, Pumpkin,

  - Chenopodiaceae : Sugar beet, Beet, Spinach,

  - Euphorbiaceae : Castor, Topioca,

  - Malvaceae : Cotton, Ladyfinger, Rosette,

  - Papilionaceae/Leguminoceae : Pea, Gram, Arhar, Groundnut, Berseem, Lathyrus, Sunhemp, Lucerne, Urd, Moong, Lentil, Soybean,

  - Linaceae : Linseed,

  - Pedaliaceae : Sesamum (Til),

  - Poaceae/Graminae : Cereals, Millets, Grasses, Sugarcane, Napier, Oat,

  - Polygonaceae : Buckwheat,

  - Solanaceae : Potato, Tobacco, Tomato, Chili, Brinjal.

- Classification According to Botany:

  - Cereal Crops:

    - Rice : Oryza sativa L.,

    - Wheat : Triticum aestivum L.,

    - Maize : Zea mays L.,

    - Bajra/Pearl millet : Pennisetum typhoides/P. glaucum L.,

    - Sorghum/Jowar : Sorghum bicolor/S. vulgare L. Moench,

    - Barley : Hordeum vulgare L.,

    - Triticale : Secale cereal,

    - Buckwheat/Pseudo cereal : Fagopyrum esculentum.

- Millet Crops:
  - Cheena/Proso millet : Panicum miliacearum,
  - Foxtail/Italian/Jerman millet/Kakun : Seteria italica L. Beauv,
  - Kodo/Coarsest millet : Paspulum scrobiculatum L.,
  - Little millet : Panicum sumatrense,
  - Madua/Ragi/Finger millet : Eleusine coracana Gaertn,
  - Sawan/Barnyard millet : Echinochloafrumentance L.

- Pulse Crops:
  - Gram/Chickpea/Bengal gram : Cicer aeritinum L.,
  - Field pea/Grain pea : Pisum sativum var. arvense,
  - Arhar/Pigeonpea/Red gram : Cajanus cajan L. Millsp.,
  - Soybean : Glycine max L. Merril,
  - Black gram/Urdbean : Vigna mungo/Phaseolus mungo L. Hepper,
  - Green gram/Moong/Moongbean : Vigna radiated Phaseolus aureus L. Wilczek,
  - French bean/Rajmash : Phaseolus vulgaris,
  - Indian cowpea/Lobia : Vigna unguiculata/V. sinensis L.,
  - Lentil : Lens esculantum/L. culinaris Moench,
  - Lathyrus/Chickling pea/Grasspea : Lathyrus sativus,
  - Mothbean : Vigna/Phaseolus aconotifolia,
  - Horse gram/Kulthi : Macrotyloma uniflorum.

- Edible Oilseed Crops:
  - Groundnut/Peanut/Monkey nut : Arachis hypogea L.,
  - Sunflower : Helianthus annus L.,
  - Safflower : Carthamus tinctorius L.,
  - Rapeseed and Mustard : Brassica spp. L.,
  - Sesamum/Til : Sesamum indicum L.,

- ○ Niger : Guzotta abssicinia,
- ○ Linseed/Flex : Linnum ussitatisimum L.,

- Non-edible Oilseed Crops:
  - ○ Castor : Ricinus communis L.

- Fiber Crops:
  - ○ Cotton : Gossipium spp.,
  - ○ Jute/Tita pat : Corchorus capsularis,
  - ○ Sunhemp : Crotolaria juncea L.

- Forage Crops:
  - ○ Berseem : Trifolium alexandrinum L.,
  - ○ Lucerne/Alfalfa : Medicago sativa L.,
  - ○ Oat : Avena sativa L.,
  - ○ Napier grass : Pennisetum purpureum L.,
  - ○ Clusterbean/Gaur : Cymopsis tetragonalaba L.

- Sugar Crops:
  - ○ Sugarcane/Cane : Saccharum officinarum L.,
  - ○ Sugar beet : Beta vulgaris L.

- Tuber Crops:
  - ○ Potato : Solanum tuberosum L.,
  - ○ Tapioca : Manihot utilissima.

- Stimulate Crops:
  - ○ Tobacco : Nicotiana spp.,
  - ○ Opium : Papaver somniferum.

- Medicinal Crops:
  - ○ Safed musli : Chlorophytum borivilianum,
  - ○ Ashwagandha/Winter cherry : Withania somnifera,

- ○ Rouvolfia/Sarpagandha : Rouvolfia serpentina,

- ○ Isabgol : Plantago ovata,

- ○ Butch : Acorus calamus,

- ○ Bramhi : Bacopa morriei,

- ○ Nux vomica : Strychnos nuxvomica.

- Aromatic Crops:

  - ○ Lemon grass : Cymbopoganflexuasus,

  - ○ Mentha/Mint : Menthe arvensis,

  - ○ Khus/Vetivar : Vetiveria zizanoides,

  - ○ Citronella : Cymbopogan ivinterianus,

  - ○ Tulsi/Basil : Ocimum sanctum.

- Classification Based on Special Purpose:

  - ○ Arable Crops: Crops which are cultivated on ploughed land. They are annual crops and include cereals, root crops, tobacco, sugarcane, maize and potatoes.

  - ○ Aromatic Crops: The crop/plants contain odoriferous and volatile substances, which occur as essential oils, gum exudates; balsam and oleoresin in one or more part of plant, viz. wood, bark, foliage, flower and fruit are called Aromatic plants.

  - ○ Alley Crops: Alley crops or hedgerow intercrops is grown under an agro forestry practice in which perennial, preferably leguminous trees or shrubs are grown simultaneously with arable crop. The trees, managed as hedgerows, are grown in wide rows and the crop is planted in the interspaced or 'alley' between the tree rows, e.g. Sweet potato, Black gram, Turmeric & Ginger are grown in the passages formed by the rows of Eucalyptus, Subaru, Cassia etc.

  - ○ Augment Crops: Such crops are grown to supplement the yield of the main crops, e.g. Japanese mustard with be seem.

  - ○ Avenue Crops: Such crops are grown along farm roads and fences, e.g. Pigeon pea, Glyricidia sisal etc.

  - ○ Border/Guard Crops: Such crops protect another crop from trespassing of animals or restrict the speed of wind and are mainly grown as border, e.g.

Safflower (thorny oilseed crop) is planted around the field of gram.

o   **Cash Crops:** A crop, such as tobacco, grown for direct sale rather than for livestock feed or a crop grown by a farmer primarily for sale to others rather than for his or her own use, e.g. Sugarcane, Cotton, Jute, Tobacco etc.

o   **Catch/Contingent Crops:** Such crops are cultivated to catch the forthcoming season when main crop is failed, e.g. Linseed, Toria, Urd, Moong, Cowpea etc.

o   **Contour Crops:** Crops are grown on or along the contour line to protect the land from soil erosion, e.g. Marvels grass etc.

o   **Cover Crops:** A close-growing crop grown primarily to improve and protect the soil from erosion through their ground covering foliage and/or root mats between periods of regular crop production, e.g. Labia, Groundnut, Urd, Sweet Potato, Methyl etc.

o   **Complementary Crops:** Both main and intercrop is benefited to each other, e.g. Jowar, Labia etc.

o   **Competitive Crops:** Such crops compete to each other and are unsuitable for inter-cropping, e.g. Two cereals.

o   **Exhaustive Crops:** These crops leave the field exhaustive after growing, e.g. Cereals (Rice).

o   **Energy Crops:** An energy crop is a plant grown as a low cost and low maintenance harvest used to make befouls, or directly exploited for its energy content, e.g, Sugarcane, Potato, Maize, Topioca etc.

o   **Fouling Crops:** Such crops whose culture practices allow the infestation of weeds intensively, e.g. Direct seeded upland rice.

o   **Ley Crops:** Any crop or combination of crops is grown for grazing or harvesting for immediate or future feeding to livestock, e.g. Berseem, Mustard etc.

o   **Medicinal Crops:** The crop/plant contains alkaloids, glycosides, steroids or other groups of compounds of medicinal value, which is used commercially, such plants are called Medicinal plants.

o   **Mulch Crops:** Such crops are grown to conserve the soil moisture through their ground covering foliage, e.g. Cowpea.

o   **Nurse Crops:** A crop of trees (nurse trees) shrubs or other plants introduced to foster or nourishment of another crops by i.e. shading it protecting it from frost, insulation or wind. The widest use of nurse corps is in the establishment of leguminaceous plants such as alfalfa, clover e.g. Sun hemp in sugarcane, Jowar in cowpea, Rai in pea.

○ Paira/Uteri Crops: The seed of succeeding crops like lentil, gram, pea, lathy us, bireme, linseed etc. is sown broadcast at 10 to 15 days before harvesting rice crop. This practice saves time; money (to be spent on land preparation etc.) utilizes residual fertility. This practice is common in both upland and lowland rice culture.

○ Paired row Crops: Generally the third row of crop is removed or growing of crop in pair row and the third row is escaped with an object to conserve the soil moisture in dry land areas.

○ Restorative Crops: Restorative crops are crops that help in maintaining the fertility of the soil, e.g. Pulses, Legumes etc.

○ Silage Crops: Such crops like corn, legumes, and grasses that have been harvested at early maturity, finely chopped, packed tightly to exclude air, and stored in tower silos, pits, or trenches for properly fermentation which is used as animal feed during lean period or off season, e.g. Maize, Cowpea, Jowar etc.

○ Smother Crops: Smother crops are specialized cover crops being ability to suppress weeds by providing dense foliage and quick growing ability, e.g. Buckwheat, Mustard, Cowpea, Urd etc.

○ Stimulate Crops: Stimulate crops stimulate the human body, e.g. Tobacco, Opium etc.

○ Supplementary Crops: Such cops are neither complementary nor competitive, e.g. Maize, Cucurbits etc.

○ Trap Crops: Trap crops are grown to protect the main cash crop from a certain pest or several pests. These crops are planted in completely surrounding the main cash crop and prevent pest attack from all sides of the field through attracting the pest, e.g. Cotton red bug trapped by ladyfinger around cotton.

○ Truck Crops: Growing one or more vegetable crops on a large scale for fresh shipment to distant markets. Most important truck crops are potato, tomatoes, lettuce, melons, beets, broccoli, celery, radishes, onions, cabbage, and strawberries.

• Classification Based on Cultural Method/Water:

○ Rain Fed: Crops grow only on rain water, e.g Jowar, Bajra, Mung etc.

- ○ Irrigated Crops: Crops grown with the help of irrigation water, e.g,. Chili, Sugarcane, Banana, Papaya etc.

- Classification Based on Root System:

  - ○ Tap root system: The main root goes deep into the soil, e.g. Tur, Grape, Cotton etc.

  - ○ Adventitious/Fiber rooted: The crops whose roots are fibrous shallow & spreading into the soil, e.g. Cereal crops, Wheat, Rice etc.

- Classification Based on Number of Cotyledon:

  - ○ Monocots or Monocotyledons: Having one cotyledon in the seed, e.g. All cereals & millets.

  - ○ Dicots or Dicotyledonous: Crops having two cotyledons in the seed, e.g. All legumes & pulses.

Cereal Crop

Cereal crops are mostly grasses cultivated for their edible seeds (actually a fruit called a caryopsis). Cereal grains are grown in greater quantities worldwide than any other type of crop and provide more food energy to the human race than any other crop. Wheat and rice are the most important crops worldwide as they account for over 50% of the world's cereal production. In the UK, wheat is the cereal most commonly used for the manufacture of food products, although many other types of cereals (e.g. maize and barley) are used. The starchy carbohydrates which are provided by cereals are essential in human nutrition. Rice is a staple diet for half the world's population, the remaining half cultivating the other cereals depending on climate and soil. Nowadays, cereals provide a very significant proportion of both human and animal diets despite the fact that most grains are, to a greater or lesser extent, deficient in a number of essential nutrients. A primary problem is the low level of essential amino acids such as lysine, methionine and threonine in the major cereal storage proteins.

The seven principal cereals grown in the world are wheat, maize, rice, barley, oats, rye and sorghum. Wheat became very popular because of the bread produced. In Europe, rye was very popular for bread-making. Barley was also introduced from Europe and was commonly grown for its high food value for both animals and humans. Oats had a lower food value than barley but they grow well in the damper climate.

Wheat (Triticum spp.) is a grass cultivated worldwide. Wheat is usually ground to flour which is used to produce a wide range of products. The type of flour produced differs according to the rate of extraction. Wheat grain is a staple food used to make flour, livestock feed and for fermentation to make alcohol. Wheat can be fermented to make beer. The husk can be separated and ground into bran. Wheat is also planted as a forage crop for livestock and the straw can be used as a ruminant feed component or construction material.

Maize (or corn, Zea mays) is a cereal grain that was domesticated in Mesoamerica. It is called corn in the USA, Canada and Australia, but in other countries that term may refer to other cereal grains. Hybrid maize is favored by farmers over conventional varieties for its high grain yield. It may be processed to make many different ingredients (e.g. high fructose corn syrup which can be used as an alternative to sucrose derived from sugar cane and sugar beet) and food products. Its germ is rich in oil and can be refined to produce corn oil. Finally, maize is one of the first crops for which genetically modified varieties make up a significant proportion of the total harvest.

Rice is a dietary staple of more than half of the world's human population (most of Asia and Latin America), making it the most consumed cereal grain. Rice cultivation is well suited to countries and regions with low labor costs and high rainfall, as it is very labor-intensive to cultivate and requires plenty of water for irrigation. However, it can be grown practically anywhere, even on steep hillsides. Rice is the world's third largest crop, behind maize (corn) and wheat. Brown rice has its outer husk removed and white rice is milled and polished further to remove the bran and germ.

Barley (Hordeum vulgare) is a major food and animal feed crop, a member of the grass family Poaceae. Barley is mainly sold as pearl barley, which is the whole grain with its husk removed. It is also used in bread (as flour) and ground as porridge in some countries. Barley is used mainly for malting. Malting is the process where the barley grain is germinated thus producing enzymes which convert its starch reserves to sugars, mainly maltose. In animal feed compositions based on barley grain, industrially derived lysine and threonine are added to obtain a balanced nutritional diet. However, the essential amino acids must be added at additional cost. Other non-essential amino acids, such as glutamine and proline, are present in excess in the major storage proteins and create a different problem. These amino acids, when digested by the animal, release non-utilizable nitrogen. This nitrogen is excreted in the urine, creating a significant environmental load, especially on and around pig farms. In 2004, barley ranked fourth in area of cultivation of cereal crops in the world (570 000 km2) FAOSTAT.

Oat (Avena sativa) is a species of cereal grain and the seeds of this plant. It is used for food for people and as fodder for animals, especially poultry and horses. Oat straw is used as animal bedding and sometimes as animal feed. Oat straw is also used in corn dolly making and it is the favorite filling for home-made lace pillows.

Rye is also a crop that has declined in popularity and today is no longer grown on a substantial level. Previously grown as a poor alternative to wheat, its main function was as a fodder crop. The increased production of barley has reduced its importance. Rye contains a low amount of gluten, thus producing breads with low volume and a dense texture.

Millet Crop

Millets are a group of highly variable small-seeded grasses, widely grown around the world as cereal crops or grains for fodder and human food.

Millets are important crops in the semiarid tropics of Asia and Africa (especially in India, Mali, Nigeria, and Niger), with 97% of millet production in developing countries. The crop is favored due to its productivity and short growing season under dry, high-temperature conditions.

Millets are indigenous to many parts of the world. The most widely grown millet is pearl millet, which is an important crop in India and parts of Africa. Finger millet, proso millet, and foxtail millet are also important crop species.

Millets may have been consumed by humans for about 7,000 years and potentially had "a pivotal role in the rise of multi-crop agriculture and settled farming societies".

Millet Species

Pearl millet.

Varagu (kodo) millet.

The different species of millets are not necessarily closely related. All are members of the family Poaceae (the grasses) but can belong to different tribes or even subfamilies.

Eragrostideae tribe in the subfamily Chloridoideae:

- Eleusine coracana: Finger millet,

- Eragrostis tef: Teff – often not considered to be a millet.

Paniceae tribe in the subfamily Panicoideae:

- Genus Panicum:

    ○ Panicum miliaceum: Proso millet (Common millet, Broomcorn millet, Hog millet or White millet),

    ○ Panicum sumatrense: Little millet (also known as Samalu in Telugu and Samai in Tamil Nadu).

- Pennisetum Glaucum: Pearl millet,

- Setaria Italica: Foxtail millet, Italian millet, Panic,

- Genus Digitaria – of minor importance as crops.

    ○ Digitaria exilis: White fonio, fonio millet, and hungry rice or acha rice,

    ○ Digitaria iburua: Black fonio,

    ○ Digitaria compacta: Raishan (cultivated in the Khasi Hills of northeast India),

    ○ Digitaria sanguinalis: Polish millet.

- Genus Echinochloa: Collectively, the members of this genus are called barnyard grasses or barnyard millets. Other common names to identify these seeds include Jhangora, Samo seeds or Morio/Mario/Moraiaya seeds.

    ○ Echinochloa esculenta: Japanese barnyard millet,

    ○ Echinochloa frumentacea: Indian barnyard millet (also known as Sawa millet Kodisama in Andhra Pradesh and Kuthirai vaali in Tamil Nadu and Bhagar or Varai in Maharashtra),

    ○ Echinochloa stagnina: Burgu millet,

    ○ Echinochloa crus-galli: Common barnyard grass (or Cockspur grass).

- Paspalum Scrobiculatum: Kodo millet (also known as Arikelu in Andhra Pradesh and Varagu in Tamil Nadu),

- Brachiaria Deflexa: Guinea millet,

- Urochloa Ramosa: Browntop millet (also known as Korle in Karnataka).

Andropogoneae tribe also in the subfamily Panicoideae:

- Sorghum bicolor: Sorghum - usually considered a separate cereal, but sometimes known as Great millet,

- Coix lacryma-jobi: Job's tears, also known as adlay millet.

Cultivation

Pearl millet is one of the two major crops in the semiarid, impoverished, less fertile agriculture regions of Africa and southeast Asia. Millets are not only adapted to poor, droughty, and infertile soils, but they are also more reliable under these conditions than most other grain crops. This has, in part, made millet production popular, particularly in countries surrounding the Sahara in western Africa.

Millets, however, do respond to high fertility and moisture. On a per-hectare basis, millet grain production can be 2–4 times higher with use of irrigation and soil supplements. Improved breeds of millet improve their disease resistance and can significantly enhance farm yield productivity. There has been cooperation between poor countries to improve millet yields. For example, "Okashana 1", a variety developed in India from a natural-growing millet variety in Burkina Faso, doubled yields. This breed was selected for trials in Zimbabwe. From there it was taken to Namibia, where it was released in 1990 and enthusiastically adopted by farmers. Okashana 1 became the most popular variety in Namibia, the only non-Sahelian country where pearl millet – locally known as *mahangu* – is the dominant food staple for consumers. "Okashana 1" was then introduced to Chad. The breed has significantly enhanced yields in Mauritania and Benin.

Production

In 2016, global production of millet was 28.4 million tonnes, led by India with 36% of the world total. Niger also had significant production.

Alcoholic Beverages

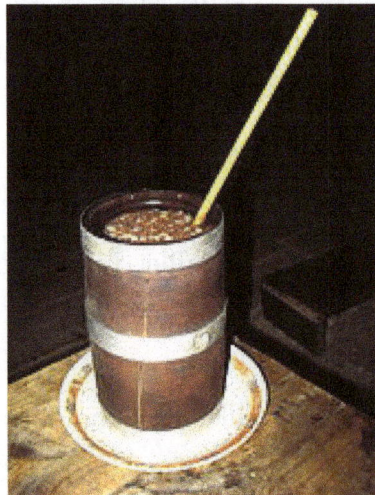

Tongba, a millet-based alcoholic brew.

In India, various alcoholic beverages are produced from millets. Millet is also the base ingredient for the distilled liquor *rakshi*.

As a Food Source

Awaokoshi, candied millet puffs, are a specialty of Osaka, Japan. This millet confection tradition began when it was presented to Sugawara no Michizane when he stopped in Naniwa during the early Heian period, about 1000 years ago.

Bánh đa kê, a specialty snack in Hanoi.

Millets are major food sources in arid and semiarid regions of the world, and feature in the traditional cuisine of many others. In western India, sorghum (called *jowar*, *jola*, *jonnalu*, *jwaarie*, or *jondhahlaa* in Gujarati, Kannada, Telugu, Hindi and Marathi languages, respectively; *mutthaari*, *kora*, or *panjappullu* in Malayalam; or *cholam* in Tamil) has been commonly used with millet flour (called *jowari* in western India) for hundreds of years to make the local staple, hand-rolled (that is, made without a rolling pin) flat bread (*rotla* in Gujarati, *bhakri* in Marathi, or *roti* in other languages). Another cereal grain popularly used in rural areas and by poor people to consume as a staple in the form of *roti*. Other millets such as *ragi* (finger millet) in Karnataka, *naachanie* in Maharashtra, or *kezhvaragu* in Tamil, "ragulu" in Telugu, with the popular *ragi rotti* and *Ragi mudde* is a popular meal in Karnataka. Ragi, as it is popularly known, is dark in color like rye, but rougher in texture.

Millet porridge is a traditional food in Russian, German, and Chinese cuisines. In Russia, it is eaten sweet (with milk and sugar added at the end of the cooking process) or savoury with meat or vegetable stews. In China, it is eaten without milk or sugar, frequently with beans, sweet potato, and various types of squash. In Germany, it is also eaten sweet, boiled in water with apples added during the boiling process and honey added during the cooling process.

Millet is also the main ingredient in a Vietnamese sweet snack called bánh đa kê. It contains a layer of smashed millet and mungbean topped with sliced dried coconut meat wrapped in a crunchy rice cake. It is a specialty of Hanoi.

Per capita consumption of millets as food varies in different parts of the world with consumption being the highest in Western Africa. In the Sahel region, millet is estimated to account for about 35 percent of total cereal food consumption in Burkina Faso, Chad and the Gambia. In Mali and Senegal, millets constitute roughly 40 percent of total cereal food consumption per capita, while in Niger and arid Namibia it is over 65 percent. Other countries in Africa where millets are a significant food source include Ethiopia, Nigeria and Uganda. Millet is also an important food item for the population living in the drier parts of many other countries, especially in eastern and central Africa, and in the northern coastal countries of western Africa. In developing countries outside Africa, millet has local significance as a food in parts of some countries, such as China, India, Burma and North Korea.

The use of millets as food fell between the 1970s and the 2000s, both in urban and rural areas, as developing countries such as India have experienced rapid economic growth and witnessed a significant increase in per capita consumption of other cereals.

People affected by gluten-related disorders, such as coeliac disease, non-celiac gluten sensitivity and wheat allergy sufferers, who need a gluten-free diet, can replace gluten-containing cereals in their diets with millet. Nevertheless, while millet does not contain gluten, its grains and flour may be contaminated with gluten-containing cereals. It is a common ingredient in seeded bread. Millets are also used as bird and animal feed.

Grazing Millet

In addition to being used for seed, millet is also used as a grazing forage crop. Instead of letting the plant reach maturity, it can be grazed by stock and is commonly used for sheep and cattle.

Millet is a C4 plant, which means that it has good water-use efficiency and utilizes high temperature and is therefore a summer crop. A C4 plant uses a different enzyme in photosynthesis from C3 plants, and this is why it improves water efficiency.

In southern Australia millet is used as a summer quality pasture, utilizing warm temperatures and summer storms. Millet is frost-sensitive and is sown after the frost period, once soil temperature has stabilised at 14 °C or higher. It is sown at a shallow depth.

Millet grows rapidly and can be grazed 5–7 weeks after sowing, when it is 20–30 cm high. The highest feed value is from the young green leaf and shoots. The plant can quickly come to head, so it must be managed accordingly because as the plant matures, the value and palatability of feed reduces.

The Japanese millets (*Echinochloa esculenta*) are considered the best for grazing and in particular Shirohie, a new variety of Japanese millet, is the best suited variety for grazing. This is due to a number of factors: it gives better regrowth and is later to mature compared to other Japanese millets; it is cheap – cost of seed is $2–$3 per kg, and sowing rates are around 10 kg per hectare for dryland production; it is quick to establish, can be grazed early, and is suitable for both sheep and cattle.

Compared to forage sorghum, which is grown as an alternative grazing forage, animals gain weight faster on millet, and it has better hay or silage potential, although it produces less dry matter. Lambs do better on millet compared to sorghum. Millet does not contain prussic acid, which can be in sorghum. Prussic acid poisons animals by inhibiting oxygen utilisation by the cells and is transported in the blood around the body — ultimately the animal will die from asphyxia. There is no need for additional feed supplements such as Sulphur or salt blocks with millet.

The rapid growth of millet as a grazing crop allows flexibility in its use. Farmers can wait until sufficient late spring/summer moisture is present and then make use of it. It is ideally suited to irrigation where livestock finishing is required.

Nutrition

In a 100 gram serving, raw millet provides 378 calories and is a rich source (20% or more of the Daily Value, DV) of protein, dietary fiber, several B vitamins and numerous dietary minerals, especially manganese at 76% DV. Raw millet is 9% water, 73% carbohydrates, 4% fat and 11% protein.

Comparison with other Major Staple Foods

The following table shows the nutrient content of millet compared to major staple foods in a raw form. Raw forms, however, are not edible and cannot be fully digested. These must be prepared and cooked as appropriate for human consumption. In processed and cooked form, the relative nutritional and antinutritional contents of each of these grains is remarkably different from that of raw forms reported in this table. The nutritional value in the cooked form depends on the cooking method.

Nutrient profile comparison of millet with other food staples							
Component (per 100 g portion, raw grain)	Cassava	Wheat	Rice	Maize	Sorghum millet	Proso millet	Kodo millet
Water (g)	60	13.1	12	76	9.2	8.7	
Energy (kj)	667	1368	1527	360	1418	1582	1462
Protein (g)	1.4	12.6	7	3	11.3	11	9.94
Fat (g)	0.3	1.5	1	1	3.3	4.2	3.03
Carbohydrates (g)	38	71.2	79	19	75	73	63.82
Fiber (g)	1.8	1.2	1	3	6.3	8.5	8.2
Sugars (g)	1.7	0.4	>0.1	3	1.9		
Iron (mg)	0.27	3.2	0.8	0.5	4.4	3	3.17
Manganese (mg)	0.4	3.9	1.1	0.2	<0.1	1.6	
Calcium (mg)	16	29	28	2	28	8	32.33
Magnesium (mg)	21	126	25	37	<120	114	
Phosphorus (mg)	27	288	115	89	287	285	300
Potassium (mg)	271	363	115	270	350	195	
Zinc (mg)	0.3	2.6	1.1	0.5	<1	1.7	32.7
Pantothenic acid (mg)	0.1	0.9	1.0	0.7	<0.9	0.8	
Vitb6 (mg)	0.1	0.3	0.2	0.1	<0.3	0.4	
Folate (µg)	27	38	8	42	<25	85	
Thiamin (mg)	0.1	0.38	0.1	0.2	0.2	0.4	0.15
Riboflavin (mg)	<0.1	0.1	>0.1	0.1	0.1	0.3	2.0
Niacin (mg)	0.9	5.5	1.6	1.8	2.9		0.09

Nutrient content of various raw millets with comparison to quinoa, teff, fonio, rice and wheat					
Crop / nutrient	Protein (g)	Fiber (g)	Minerals (g)	Iron (mg)	Calcium (mg)
Sorghum	10	4	1.6	2.6	54
Pearl millet	10.6	1.3	2.3	16.9	38
Finger millet	7.3	3.6	2.7	3.9	344
Foxtail millet	12.3	8	3.3	2.8	31
Proso millet	12.5	2.2	1.9	0.8	14
Kodo millet	8.3	9	2.6	0.5	27
Little millet	7.7	7.6	1.5	9.3	17
Barnyard millet	11.2	10.1	4.4	15.2	11
Brown top millet	11.5	12.5	4.2	0.65	0.01
Quinoa	14.1	7	*	4.6	47
Teff	13	8	0.85	7.6	180
Fonio	11	11.3	5.31	84.8	18
Rice	6.8	0.2	0.6	0.7	10
Wheat	11.8	1.2	1.5	5.3	41

Perennial Crop

Perennial crops are crops that – unlike annual crops – don't need to be replanted each year. After harvest, they automatically grow back. By eliminating replanting, perennial cropping can reduce topsoil losses due to erosion, increase biological carbon sequestration due to reduced soil-disturbing tillage, and greatly reduce waterway pollution through agricultural runoff due to less nitrogen input.

Mechanisms

- Erosion control: Because plant materials (stems, crowns, etc.) can remain in place year-round, topsoil erosion due to wind and rainfall/irrigation is reduced.

- Water-use efficiency: Because these crops tend to be deeper and more fibrously-rooted than their annual counterparts, they are able to hold onto soil moisture more efficiently, while filtering pollutants (e.g. excess nitrogen) traveling to groundwater sources.

- Nutrient cycling efficiency: Because perennials more efficiently take up nutrients as a result of their extensive root systems, reduced amounts of nutrients need to be supplemented, lowering production costs while reducing possible excess sources of fertilizer runoff.

- Light interception efficiency: Earlier canopy development and longer green leaf duration increase the seasonal light interception efficiency of perennials, an important factor in plant productivity.

Example Crops

- Miscanthus giganteus: A perennial crop with high yields and high GHG mitigation potential.

- Perennial sunflower: A perennial oil and seedcrop developed through back-crossing genes with wild sunflower.

- Perennial grain: More extensive root systems allow for more efficient water and nutrient uptake, while reducing erosion due to rain and wind year-round.

- Perennial rice: Currently in the development stage using similar methods to those used in producing the perennialized sunflower, perennial rice promises to reduce deforestation through increases in production efficiency by keeping cleared land out of the fallow stage for long periods of time.

Cash Crop

A cash crop or profit crop is an agricultural crop which is grown to sell for profit. It is typically purchased by parties separate from a farm. The term is used to differentiate marketed crops from subsistence crops, which are those fed to the producer's own livestock or grown as food for the producer's family. In earlier times cash crops were usually only a small (but vital) part of a farm's total yield, while today, especially in developed countries, almost all crops are mainly grown for revenue. In the least developed countries, cash crops are usually crops which attract demand in more developed nations, and hence have some export value.

Prices for major cash crops are set in commodity markets with global scope, with some local variation (termed as "basis") based on freight costs and local supply and demand balance. A consequence of this is that a nation, region, or individual producer relying on such a crop may suffer low prices should a bumper crop elsewhere lead to excess supply on the global markets. This system has been criticized by traditional farmers. Coffee is an example of a product that has been susceptible to significant commodity futures price variations.

Globalization

Issues involving subsidies and trade barriers on such crops have become controversial in discussions of globalization. Many developing countries take the position that the current international trade system is unfair because it has caused tariffs to be lowered in industrial goods while allowing for low tariffs and agricultural subsidies for agricultural goods. This makes it difficult for a developing nation to export its goods overseas, and forces developing nations to compete with imported goods which are exported from developed nations at artificially low prices. The practice of exporting at artificially low prices is known as dumping, and is illegal in most nations. Controversy over this issue led to the collapse of the Cancún trade talks in 2003, when the Group of 22 refused to consider agenda items proposed by the European Union unless the issue of agricultural subsidies was addressed.

Per Climate Zones

Arctic

The Arctic climate is generally not conducive for the cultivation of cash crops. However, one potential cash crop for the Arctic is *Rhodiola rosea*, a hardy plant used as a medicinal herb that grows in the Arctic. There is currently consumer demand for the plant, but the available supply is less than the demand.

Temperate

Cash crops grown in regions with a temperate climate include many cereals (wheat, rye, corn, barley, oats), oil-yielding crops (e.g. grapeseed, mustard seeds), vegetables (e.g. potatoes), tree fruit or top fruit (e.g. apples, cherries) and soft fruit (e.g. strawberries, raspberries).

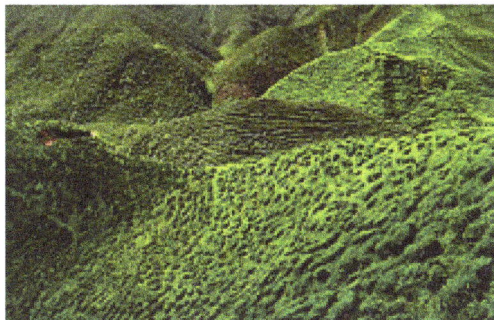

A tea plantation in the Cameron Highlands in Malaysia.

Subtropical

In regions with a subtropical climate, oil-yielding crops (e.g. soybeans) and some vegetables and herbs are the predominant cash crops.

Tropical

In regions with a tropical climate, coffee, cocoa, sugar cane, bananas, oranges, cotton and jute (a soft, shiny vegetable fiber that can be spun into coarse, strong threads), are common cash crops. The oil palm is a tropical palm tree, and the fruit from it is used to make palm oil.

Global Cash Crops

Coconut palms are cultivated in more than 80 countries of the world, with a total production of 61 million tonnes per year. The oil and milk derived from it are commonly used in cooking and frying; coconut oil is also widely used in soaps and cosmetics.

Sustainability of Cash Crops

Approximately 70% of the world's food is produced by 500 million smallholder farmers. For their livelihood they depend on the production of cash crops, basic commodities that are hard to differentiate in the market. The great majority (80%) of the world's farms measure 2 hectares or less. These smallholder farmers are mainly found in developing countries and are often unorganized, illiterate or enjoyed only basic education. Smallholder farmers have little bargaining power and incomes are low, leading to a situation in which they cannot invest much in upscaling their businesses. In general, farmers lack access to agricultural inputs and finance, and do not have enough knowledge on good agricultural and business practices. These high level problems are in many cases threatening the future of agricultural sectors and theories start evolving on how to secure a sustainable future for agriculture. Sustainable market transformations are initiated in which industry leaders work together in a pre-competitive environment to change market conditions. Sustainable intensification focuses on facilitating entrepreneurial farmers. To stimulate farm investment, projects on access to finance for agriculture are also popping up. One example is the SCOPE methodology, an assessment tool that measures the management maturity and professionalism of producer organizations as to give financing organizations better insights in the risks involved in financing. Currently agricultural finance is always considered risky and avoided by financial institutions.

Black Market Cash Crops

Cannabis has been termed as a cash crop.

Coca, opium poppies and cannabis are significant black market cash crops, the prevalence of which varies. In the United States, cannabis is considered by some to be the most valuable cash crop. In 2006, it was reported in a study by Jon Gettman, a marijuana policy researcher, that in contrast to government figures for legal crops such as corn and wheat and using the study's projections for U.S. cannabis production at that time, cannabis was cited as "the top cash crop in 12 states and among the top three cash crops in 30". The study also estimated cannabis production at the time to be valued at $35.8 billion USD, which exceeded the combined value of corn at $23.3 billion and wheat at $7.5 billion.

Cover Crop

A cover crop is planted to manage soil erosion, soil fertility, soil quality, water, weeds, pests, diseases, biodiversity and wildlife in an *agroecosystem*—an ecological system managed and shaped by humans. Cover crops may be an off-season crop planted after harvesting the cash crop. The cover crop may grow over winter.

Soil Erosion

Although cover crops can perform multiple functions in an agroecosystem simultaneously, they are often grown for the sole purpose of preventing soil erosion. Soil erosion is a process that can irreparably reduce the productive capacity of an agroecosystem. Dense cover crop stands physically slow down the velocity of rainfall before it contacts the soil surface, preventing soil splashing and erosive surface runoff. Additionally, vast cover crop root networks help anchor the soil in place and increase soil porosity, creating suitable habitat networks for soil macrofauna. It keeps the enrichment of the soil good for the next few years.

Soil Fertility Management

One of the primary uses of cover crops is to increase soil fertility. These types of cover crops are referred to as "green manure." They are used to manage a range of soil macronutrients and micronutrients. Of the various nutrients, the impact that cover crops have on nitrogen management has received the most attention from researchers and farmers, because nitrogen is often the most limiting nutrient in crop production.

Often, green manure crops are grown for a specific period, and then plowed under before reaching full maturity in order to improve soil fertility and quality. Also the stalks left block the soil from being eroded.

Green manure crops are commonly leguminous, meaning they are part of the Fabaceae (pea) family. This family is unique in that all of the species in it set pods, such as bean, lentil, lupins and alfalfa. Leguminous cover crops are typically high in nitrogen and can often provide the required quantity of nitrogen for crop production. In conventional farming, this nitrogen is typically applied in chemical fertilizer form. This quality of cover crops is called fertilizer replacement value.

Another quality unique to leguminous cover crops is that they form symbiotic relationships with the rhizobial bacteria that reside in legume root nodules. Lupins is nodulated by the soil microorganism *Bradyrhizobium* sp. (Lupinus). Bradyrhizobia are encountered as microsymbionts in other leguminous crops (*Argyrolobium, Lotus, Ornithopus, Acacia, Lupinus*) of Mediterranean origin. These bacteria convert biologically unavailable atmospheric nitrogen gas (N_2) to biologically available ammonium (NH^+_4) through the process of biological nitrogen fixation.

Prior to the advent of the Haber-Bosch process, an energy-intensive method developed to carry out industrial nitrogen fixation and create chemical nitrogen fertilizer, most nitrogen introduced to ecosystems arose through biological nitrogen fixation. Some scientists believe that widespread biological nitrogen fixation, achieved mainly through the use of cover crops, is the only alternative to industrial nitrogen fixation in the effort to maintain or increase future food production levels. Industrial nitrogen fixation has been criticized as an unsustainable source of nitrogen for food production due to its reliance on fossil fuel energy and the environmental impacts associated with chemical nitrogen fertilizer use in agriculture. Such widespread environmental impacts include nitrogen fertilizer losses into waterways, which can lead to eutrophication (nutrient loading) and ensuing hypoxia (oxygen depletion) of large bodies of water.

An example of this lies in the Mississippi Valley Basin, where years of fertilizer nitrogen loading into the watershed from agricultural production have resulted in a hypoxic "dead zone" off the Gulf of Mexico the size of New Jersey. The ecological complexity of marine life in this zone has been diminishing as a consequence.

As well as bringing nitrogen into agroecosystems through biological nitrogen fixation, types of cover crops known as "catch crops" are used to retain and recycle soil nitrogen already present. The catch crops take up surplus nitrogen remaining from fertilization of the previous crop, preventing it from being lost through leaching, or gaseous denitrification or volatilization.

Catch crops are typically fast-growing annual cereal species adapted to scavenge available nitrogen efficiently from the soil. The nitrogen tied up in catch crop biomass is released back into the soil once the catch crop is incorporated as a green manure or otherwise begins to decompose.

An example of green manure use comes from Nigeria, where the cover crop *Mucuna pruriens* (velvet bean) has been found to increase the availability of phosphorus in soil after a farmer applies rock phosphate.

Soil Quality Management

Cover crops can also improve soil quality by increasing soil organic matter levels through the input of cover crop biomass over time. Increased soil organic matter enhances soil structure, as well as the water and nutrient holding and buffering capacity of soil. It can also lead to increased soil carbon sequestration, which has been promoted as a strategy to help offset the rise in atmospheric carbon dioxide levels.

Soil quality is managed to produce optimum circumstances for crops to flourish. The principal factors of soil quality are soil salination, pH, microorganism balance and the prevention of soil contamination.

Water Management

By reducing soil erosion, cover crops often also reduce both the rate and quantity of water that drains off the field, which would normally pose environmental risks to waterways and ecosystems downstream. Cover crop biomass acts as a physical barrier between rainfall and the soil surface, allowing raindrops to steadily trickle down through the soil profile. Also, as stated above, cover crop root growth results in the formation of soil pores, which in addition to enhancing soil macrofauna habitat provides pathways for water to filter through the soil profile rather than draining off the field as surface flow. With increased water infiltration, the potential for soil water storage and the recharging of aquifers can be improved.

Just before cover crops are killed (by such practices including mowing, tilling, discing, rolling, or herbicide application) they contain a large amount of moisture. When the cover crop is incorporated into the soil, or left on the soil surface, it often increases soil moisture. In agroecosystems where water for crop production is in short supply, cover crops can be used as a mulch to conserve water by shading and cooling the soil surface. This reduces evaporation of soil moisture. In other situations farmers try to dry the soil out as quickly as possible going into the planting season. Here prolonged soil moisture conservation can be problematic.

While cover crops can help to conserve water, in temperate regions (particularly in years with below average precipitation) they can draw down soil water supply in the spring, particularly if climatic growing conditions are good. In these cases, just before crop planting, farmers often face a tradeoff between the benefits of increased cover crop growth and the drawbacks of reduced soil moisture for cash crop production that season. C/N ratio is balanced with this application.

Weed Management

Cover crop.

Thick cover crop stands often compete well with weeds during the cover crop growth period, and can prevent most germinated weed seeds from completing their life cycle and reproducing. If the cover crop is flattened down on the soil surface rather than incorporated into the soil as a green manure after its growth is terminated, it can form a nearly impenetrable mat. This drastically reduces light transmittance to weed seeds, which in many cases reduces weed seed germination rates. Furthermore, even when weed seeds germinate, they often run out of stored energy for growth before building the necessary structural capacity to break through the cover crop mulch layer. This is often termed the cover crop smother effect.

Some cover crops suppress weeds both during growth and after death. During growth these cover crops compete vigorously with weeds for available space, light, and nutrients, and after death they smother the next flush of weeds by forming a mulch layer on the soil surface. For example, researchers found that when using *Melilotus officinalis* (yellow sweetclover) as a cover crop in an improved fallow system (where a fallow period is intentionally improved by any number of different management practices, including the planting of cover crops), weed biomass only constituted between 1-12% of total standing biomass at the end of the cover crop growing season. Furthermore, after cover crop termination, the yellow sweetclover residues suppressed weeds to levels 75–97% lower than in fallow (no yellow sweetclover) systems.

In addition to competition-based or physical weed suppression, certain cover crops are known to suppress weeds through allelopathy. This occurs when certain biochemical cover crop compounds are degraded that happen to be toxic to, or inhibit seed germination of, other plant species. Some well known examples of allelopathic cover crops are *Secale cereale* (rye), *Vicia villosa* (hairy vetch), *Trifolium pratense* (red clover), *Sorghum bicolor* (sorghum-sudangrass), and species in the Brassicaceae family, particularly mustards. In one study, rye cover crop residues were found to have provided between 80% and 95% control of early season broadleaf weeds when used as a mulch during the production of different cash crops such as soybean, tobacco, corn, and sunflower.

In a recent study released by the Agricultural Research Service (ARS) scientists examined how rye seeding rates and planting patterns affected cover crop production. The results show that planting more pounds per acre of rye increased the cover crop's production as well as decreased the amount of weeds. The same was true when scientists tested seeding rates on legumes and oats; a higher density of seeds planted per acre decreased the amount of weeds and increased the yield of legume and oat production. The planting patterns, which consisted of either traditional rows or grid patterns, did not seem to make a significant impact on the cover crop's production or on the weed production in either cover crop. The ARS scientists concluded that increased seeding rates could be an effective method of weed control.

Disease Management

In the same way that allelopathic properties of cover crops can suppress weeds, they can also break disease cycles and reduce populations of bacterial and fungal diseases, and parasitic nematodes. Species in the Brassicaceae family, such as mustards, have been widely shown to suppress fungal disease populations through the release of naturally occurring toxic chemicals during the degradation of glucosinolade compounds in their plant cell tissues.

Pest Management

Some cover crops are used as so-called "trap crops", to attract pests away from the crop of value and toward what the pest sees as a more favorable habitat. Trap crop areas can be established within crops, within farms, or within landscapes. In many cases the trap crop is grown during the same season as the food crop being produced. The limited area occupied by these trap crops can be treated with a pesticide once pests are drawn to the trap in large enough numbers to reduce the pest populations. In some organic systems, farmers drive over the trap crop with a large vacuum-based implement to physically pull the pests off the plants and out of the field. This system has been recommended for use to help control the lygus bugs in organic strawberry production. Another example of trap crops are nematode resistance white mustard (*Sinapis alba*) and radish (*Raphanus sativus*). They can be grown after a main (cereal) crop and trap nematodes, for example the beet cyst nematode and Columbian root knot nematode. When grown, nematodes hatch and are attracted to the roots. After entering the roots they cannot reproduce in the root due to a hypersensitive resistance reaction of the plant. Hence the nematode population is greatly reduced, by 70-99%, depending on species and cultivation time.

Other cover crops are used to attract natural predators of pests by providing elements of their habitat. This is a form of biological control known as habitat augmentation, but achieved with the use of cover crops. Findings on the relationship between cover crop presence and predator/pest population dynamics have been mixed, pointing toward the need for detailed information on specific cover crop types and management practices to best complement a given integrated pest management strategy. For example, the predator mite *Euseius tularensis* (Congdon) is known to help control the pest citrus thrips in Central California citrus orchards. Researchers found that the planting of several different leguminous cover crops (such as bell bean, woollypod vetch, New Zealand white clover, and Austrian winter pea) provided sufficient pollen as a feeding source to cause a seasonal increase in *E. tularensis* populations, which with good timing could potentially introduce enough predatory pressure to reduce pest populations of citrus thrips.

Diversity and Wildlife

Although cover crops are normally used to serve one of the above discussed purposes, they often simultaneously improve farm habitat for wildlife. The use of cover crops adds at least one more dimension of plant diversity to a cash crop rotation. Since the cover crop is typically not a crop of value, its management is usually less intensive, providing a window of "soft" human influence on the farm. This relatively "hands-off" management, combined with the increased on-farm heterogeneity created by the establishment of cover crops, increases the likelihood that a more complex trophic structure will develop to support a higher level of wildlife diversity.

In one study, researchers compared arthropod and songbird species composition and field use between conventionally and cover cropped cotton fields in the Southern United States. The cover cropped cotton fields were planted to clover, which was left to grow in between cotton rows throughout the early cotton growing season (stripcover cropping). During the migration and breeding season, they found that songbird densities were 7–20 times higher in the cotton fields with integrated clover cover crop than in the conventional cotton fields. Arthropod abundance and biomass was also higher in the clover cover cropped fields throughout much of the songbird breeding season, which was attributed to an increased supply of flower nectar from the clover. The cover crop enhanced songbird habitat by providing cover and nesting sites, and an increased food source from higher arthropod populations.

Fibre Crop

Fiber crops are field crops grown for their fibers, which are traditionally used to make paper, cloth, or rope. They are organized into 3 main groups—textile fibers (used in production of cloth), cordage fibers (used in production of rope), and filling fibers (used to stuff upholstery and mattresses). They are a type of natural fiber.

Fiber crops are characterized by having a large concentration of cellulose, which is what gives them their strength. The fibers may be chemically modified, like in viscose (used to make rayon and cellophane). In recent years, materials scientists have begun exploring further use of these fibers in composite materials. Due to cellulose being the main factor of a plant fibers strength, this is what scientists are looking to manipulate to create different types of fibers.

Fiber crops are generally harvestable after a single growing season, as distinct from trees, which are typically grown for many years before being harvested for such materials as wood pulp fiber or lacebark. In specific circumstances, fiber crops can be superior to wood pulp fiber in terms of technical performance, environmental impact or cost.

There are a number of issues regarding the use of fiber crops to make pulp. One of these is seasonal availability. While trees can be harvested continuously, many field crops are harvested once during the year and must be stored such that the crop doesn't rot over a period of many months. Considering that many pulp mills require several thousand tonnes of fiber source per day, storage of the fiber source can be a major issue. These fiber mostly found in leaf,seeds or in the body of a tree. Fiber crops are more important in daily life basis.

Botanically, the fibers harvested from many of these plants are bast fibers; the fibers come from the phloem tissue of the plant. The other fiber crop fibers are hard/leaf fibers (come from the entirety of plant vascular bundles) and surface fibers (come from plant epidermal tissue).

Fiber Sources

Before the industrialisation of the paper production the most common fiber source was recycled fibers from used textiles, called rags. The rags were from hemp, linen and cotton. A process for removing printing inks from recycled paper was invented by German jurist Justus Claproth in 1774. Today this method is called deinking. It was not until the introduction of wood pulp in 1843 that paper production was not dependent on recycled materials from ragpickers.

To have a source of fiber to utilize in production, the fiber first must be extracted from the plant. This is done in different ways depending on the fiber classification. Bast fibers are harvested through retting which is where microbes are utilized to remove soft tissues from the plant and only the useful fibrous material remains. Hard fibers are harvested mainly through decortication which is where the non-fibrous tissues are removed by hand or machine. Lastly, surface fibers are harvested through ginning which is where a machine removes the fibers from other plant material.

Fiber Crops

- Bast fibers (Stem-skin fibers):
  - Esparto, a fiber from a grass,
  - Jute, widely used, it is the cheapest fiber after cotton,
  - Flax, produces linen,
  - Indian hemp, the Dogbane used by Native Americans,
  - Hemp, a soft, strong fiber, edible seeds,
  - Hoopvine, also used for barrel hoops and baskets, edible leaves, medicine,
  - Kenaf, the interior of the plant stem is used for its fiber. Edible leaves,
  - Beans, an edible seed, typically kidney-shaped, growing in long pods on certain leguminous plants,
  - Linden Bast,
  - Nettles,
  - Ramie, a nettle, stronger than cotton or flax, makes "China grass cloth",
  - Papyrus, a pith fiber, akin to a bast fiber.
- Leaf fibers:
  - Abacá, a banana, producing "manila" rope from leaves,
  - Sisal, an agave,
  - Bowstring Hemp, a common house plant, also Sansevieria roxburghiana, Sansevieria hyacinthoides,
  - Henequen, an agave. A useful fiber, but not as high quality as sisal,
  - Phormium, "New Zealand Flax",

- ○ Yucca, an agave relative.
- • Seed fibers and fruit fibers:
  - ○ Coir, the fiber from the coconut husk,
  - ○ Cotton,
  - ○ Kapok,
  - ○ Milkweed, grown for the filament-like pappus in its seed pods,
  - ○ Luffa, a gourd which when mature produces a sponge-like mass of xylem, used to make loofa sponge.
- • Other fibers (Leaf, fruit, and other fibers):
  - ○ Bamboo fiber, a viscose fiber like rayon, technically a semi-synthetic fiber.

Fiber Dimensions

Source of pulp	Fiber length, mm	Fiber diameter, μm
Softwood	3.1	30
Hardwood	1.0	16
Wheat straw	1.5	13
Rice straw	1.5	9
Esparto grass	1.1	10
Reed	1.5	13
Bagasse	1.7	20
Bamboo	2.7	14
Cotton	25.0	20

Row Crop

A row crop is a crop that can be planted in rows wide enough to allow it to be tilled or otherwise cultivated by agricultural machinery, machinery tailored for the seasonal activities of row crops. Such crops are sown by drilling rather than broadcasting.

The distinction is significant in crop rotation strategies, where land is planted with row crops, commodity food grains, and sod-forming crops in a sequence meant to protect the quality of the soil while maximizing the soil's annual productivity.

Row crops are generally grown on irrigated land, and some, such as cotton, can be grown only under irrigation. During the growing season, the interrow spaces are hoed two to four times and the rows are weeded to conserve moisture and improve aeration. As a result, the soil's microbiological activity increases and mobilization of nutrients is intensified. Row crops are valuable precursors of spring grain crops, flax, and hemp. The beneficial effect of row crops extends to the second crop.

Examples of row crops include sunflower, potato, canola, dry bean, field pea, flax, safflower, buckwheat, cotton, maize, soybeans, and sugar beets.

Sustainable agriculture does not necessarily mean organically grown or locally sourced or non GMO. Sustainable agriculture at its simplest means to be able to produce the food and fiber we currently need while allowing future generations that same opportunity. This means efficiently managing our natural resources such that we don't deplete or otherwise negatively impact soil quality, water quality and water quantity. Sustainability not only includes environmental goals, but should be economically and socially sound as well.

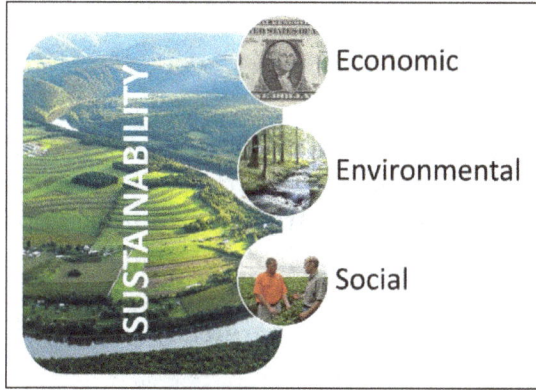

This means that an unprofitable management practice that is environmentally beneficial is not necessarily sustainable, nor is a profitable management practice that negatively impacts the environment. This fact sheet, the first in a series, will introduce general concepts and practices that are essential to sustainable row crop production.

Subsequent fact sheets will provide in-depth, localized information on each of the management practices introduced here.

Sustainable Practices for Conventional Agriculture

There are many management practices that can enable producers to reduce or more efficiently use their crop inputs while maintaining or increasing productivity, leading to economic as well as environmental benefits. These practices include, but are not limited to, the following:

No-till

No-till management is a cropping practice that minimizes soil disturbance and maintains crop residues on the soil surface as cover.

According to the National Agricultural Statistics Service, approximately 80 percent of Tennessee's production acres are no-till. This practice has substantially decreased the amount of soil loss occurring in our state.

Soil Sampling

The most basic, yet beneficial, tool for row crop nutrient management is soil testing. Soil tests help you determine which nutrients need to be applied, how much of the nutrients should be applied, and, on multiple scales, where in the field to apply the nutrients. Soil tests also inform you if lime needs to be applied to achieve an optimum soil pH. Nutrient availability and crop growth are greatly affected by soil pH.

Often, a blanket rate application leads to overapplication of crop fertilizers in some areas of the field and underapplication in others. Site-specific soil sampling, utilizing either grid or zone sampling techniques, in combination with variable rate application (VRA) of fertilizers is a beneficial and effective tool for row crop producers to manage natural soil fertility variability. By targeting fertilizer inputs to specific areas of the field, row crop producers can maximize profitability and reduce the risk of nutrient runoff or leaching.

Cover Crops

Cover crops are grasses, legumes or small grains planted in the fall and terminated before planting of the row crop in the spring. Cover crops are known to have many beneficial effects on cropping systems, such as improvements in soil health and function, erosion control and nutrient management. Cover crops improve soil quality by increasing organic matter, and, thus, microbial activity; increasing aggregate stability; and promoting infiltration. Additionally, cover crops promote nutrient cycling and provide other soil fertility benefits. Legume species have the potential to fix atmospheric nitrogen and can provide a substantial amount of nitrogen for the subsequent row crop to utilize.

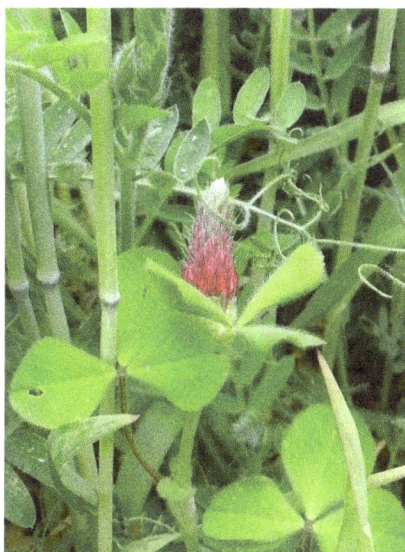

More recently, cover crops have been used for early-season weed suppression, which has the potential to further reduce chemical inputs and relieve some of the pressures of herbicide-resistant weeds.

Nutrient Stewardship

Nutrient stewardship consists of proper nutrient management practices that can not only increase row crop production efficiency, but also ease the environmental stresses related to fertilizer application and diminished water quality. The basis of this concept is to precisely manage nutrients on the farm by considering what source to apply, when to apply, where to apply and how much to apply.

A fertility program should be based on all of these aspects; for example, when selecting what source to apply, the timing, placement and rate also should be considered. There is no one-size-fits-all management strategy, and the nutrient management decisions are specific not only to individual producers, but also to individual fields.

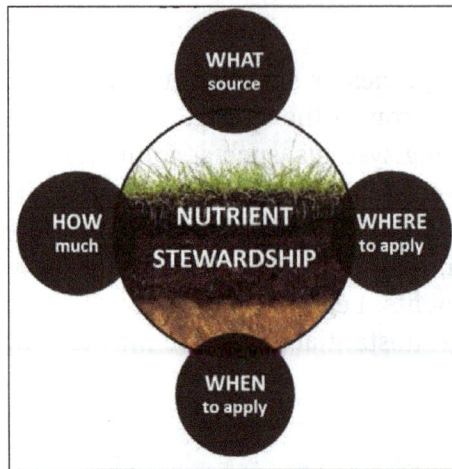

Precision Agriculture Technologies

Precision agriculture technologies dispense or apply desired amounts of crop inputs such as seeds, pesticides and fertilizers in specific areas of an agricultural field in order to maximize productivity. There are several technologies commercially available to monitor and apply these production inputs based on the Global Positioning System (GPS)/Global Navigation Satellite System (GNSS) location of farm equipment in the field and on application maps.

One example of this technology is Automatic Section Control (ASC) for row crop planters and sprayers. As the planter or sprayer travels across the field, the controller continually checks to see if sections are passing over previously planted or sprayed areas or areas that have been mapped as no coverage zones. When a planter or sprayer section passes into these areas, it is turned off automatically and turned back on when it passes back into unplanted or unsprayed areas. Reducing offtarget chemical application errors with sprayer equipped ASC can improve profitability by reducing chemical costs and losses to the environment.

Variable rate application of fertilizer allows the right rate of nutrients to be put in the right location. This technology has the potential to reduce nutrient inputs and increase nutrient use efficiency. In turn, the use of VRA may reduce the amount of nutrients that could ultimately pollute nearby streams, rivers and reservoirs.

Improving Water use Efficiency

Irrigation scheduling tools can significantly improve water use efficiency by supplying water when the crop needs it and not overapplying. Soil moisture sensors can signal you about when the soil profile is low in available water, and scheduling tools such as UT's MOIST3 (Management of Irrigation Systems in Tennessee) program can help you make decisions about when to turn your irrigation system on. MOIST uses a water balance approach, which keeps track of how much water exits a crop's root zone via plant transpiration/surface evaporation and how much enters by rainfall and irrigation. This scheduling approach enables you to efficiently maintain soil water in a range that will optimize yield.

Importance

Sustainability is not an end-goal; it should be a continuous effort among all agricultural sectors. There are many practices that can reduce environmental impacts while maintaining or improving farm profitability. Producers should consider adopting practices that can work on their farms, focusing on the well-being of our environment as well as the overall profitability of their farming operation.

Forage Crop

Forage crops are crops grown specifically to be grazed by livestock or conserved as hay or silage.

Forage crops assist in achieving production targets for attributes such as growth or weight gain and to make up seasonal short falls between feed demand and supply. They can also play an important role in maintaining ground cover, preventing erosion, accumulating nitrogen in the soil and improving land condition.

Forage crops can be an important tool for producers, provided the right crop is selected and then carefully managed during establishment and grazing to ensure maximum productivity is achieved.

Selection

A range of forage crops are available to producers. The right crop will depend on the quality and quantity of feed required to meet the enterprise production objectives as well as other factors such as soil type, climate, water availability, drainage, weeds and disease.

Careful selection will help ensure good plant growth and grazing potential.

Establishment

Soil nutrition, weed control and, sowing or seeding are important considerations when seeking to establish a forage crop.

Good soil nutrient levels are important in ensuring vigorous forage crop establishment. While the fertiliser history of a paddock can provide an insight into the likely levels of key nutrients, the only way to be sure of the soil nutrient profile is to undertake a soil test.

Conventional, minimum till and no till sowing can be used in the establishment of a forage crop. Whichever method is used, it is important to ensure good soil to seed contact as this is critical in achieving good seed germination.

Weed management is best undertaken prior to or at the time of sowing to minimise the amount of competition with the establishing crop. Whether or not the forage crop is part of an ongoing rotation will influence both the choice of crop and the options available for weed control.

Management

Once established, careful management is required to ensure that the crop is fully utilised in its most productive and nutritious phases of growth. Pests and diseases must also be managed to minimise their impact on productivity.

A well considered grazing strategy is important in maximising the productive potential of a fodder crop. Forage crops can either be grown exclusively for hay or silage production or grazed before being set aside for fodder conservation.

The timing of both grazing events and the cutting for hay or silage are critical to ensuring that the quality and quantity of conserved fodder are optimised.

Tuber Crop

Root and tuber crops are plants yielding starchy roots, tubers, rhizomes, corms, and stems. They are used mainly for human food (as such or in processed form), animal feed, and for manufacturing starch, alcohol, and fermented beverages including beer. Many of the developing world's poorest producers and the most undernourished households depend on root and tuber crops as an important source of food and nutrition. Root and tuber crops produce large quantities of energy per day, in comparison with cereals. They have been playing more and more important roles in global food and energy security. The major root and tuber crops – potato, sweet potato, cassava, and yam – occupy approximately 53.93 million hectares worldwide and produce 736.747 million tonnes annually. Individually, potato, sweet potato, cassava, and yam rank among the most important food crops worldwide in terms of annual volume of production. Potato, sweet potato, and cassava rank among the top ten food crops produced in developing countries.

Tropical root and tuber crops are important food crops serving either as subsidiary or subsistence food in different parts of the tropical belt. They are rich sources of starch besides many vitamins, minerals, etc. Although there has been some decline in their use as food, their industrial application, especially that of cassava, is making rapid advances. Cassava and to a small extent, sweet potato are used for starch extraction in countries like India, Brazil, Thailand, Indonesia, Philippines and China. Studies at different laboratories have brought to light the wide diversity in the starch characteristics of tuber crops and the possibility of using these native starches instead of chemically modified starches.

Cassava (Manihot esculenta Crantz) is a sturdy perennial crop grown in many parts of Asia, Africa and South America. The yield of the crop is normally around 20 t/ha and the starch yield forms nearly one quarter of the total yield. The starch has been studied in detail and finds use in a large range of industries.

Sweet potato is a herbaceous perennial vine and is grown extensively in the tropics and also in some parts of the USA for its tubers. The tubers have different sizes, shapes and colour. The yield varies anywhere from 8–30 t/ha and it has been possible to have three crops in a year thus giving a very high annual starch yield. The starch content in the fresh tubers varies from 12–30%.

Taro is a small herbaceous plant with large leaves found in most parts of the tropics and is very important in the pacific regions. The crop is harvested at 8–10 month stage and produces a number of cormels around a corm. The yield is 5–10 t/ha and starch content in the tubers is 12–20%.

Tannia is a large herbaceous plant grown widely for its cormels which are much larger than taro cormels. The yield is around 10– 25 t/ha and starch content in the tubers nearly 20%.

Elephant Foot Yam is grown extensively for its huge corms. It is a big perennial herb harvested after one year and the corms can weigh over 15 kg. The yield is over 20 t/ha and starch content in the tubers is approximately 20%.

Yams comprise a large genus with over 600 species from which a few are more commonly cultivated. Most of them are trailers. The tubers are harvested at 8–12 months after planting and the tubers especially those of D. alata, are very large. Some of the species produce aerial tubers also. The yield of the crop and starch yield also vary considerably among the different species and among these D. rotundata tubers have the highest starch content. In addition, there are a number of other minor tuber and root crops which contain starch, but their utilisation is limited.

African yam bean belongs to Leguminosae. It is a vigorous herbaceous climbing vine reaching 1.5–2 m in height producing pods as well as small spindle shaped tubers about 5–8 cm long similar to sweet potato. The crop is found mainly in Africa yielding up to 4 t/ha. The tubers are rich in starch (25%) but there are no studies on this starch.

Arracacha, Peruvian carrot, is a stout semi-caulescent herb, resembling celery grown in south America and parts of Africa mainly at high altitudes. The edible secondary tubers are usually 6–10 in number with a yield of 3–18 t/ha. The tubers which contain nearly 20% starch, are used as a source of edible starch.

Chinese water chestnut is a variable annual stout aquatic plant producing corms found in the Asian region. The yield is 20–40 t/ha and starch content is around 7%. In China the starch is extracted by rasping and settling.

East Indian arrowroot is a perennial herb with a tuberous rhizome producing large/small sized tubers, found in tropical areas. The tubers are similar to potato are harvested after 8–10 months and can weigh up to 1 kg. They are used in Tahiti to make 'poi' a traditional food. The tubers are rich in starch (20–30%) and it is extracted and used widely.

Giant taro is a tall succulent herbaceous plant up to 4.5 metres in height producing big corms up to 18 kg. It is grown in Asia and South America and harvested 10–12 months after planting. The tubers contain 17–25% starch.

Coleus is a small herbaceous annual 15–30 cm high found in Africa and Asia. At maturity (6 months) they yield round to oval tubers that are very prized for their delicate flavour. The yield is 7–15 t/ha and the starch content is nearly 15%.

Kudzu known as arrowroot vine, is a small perennial twining herb or shrub with elongated tuberous roots often weighing up to 40 kg and yield of 5–7 t/ha. Roots are starchy, 30–60 cm long and used as a source of edible starch in place of arrowroot starch or gelatine in many foods. The starch content is over 20% and is extracted on a small scale in Japan but the starch has not been studied.

Lotus root is a perennial aquatic herb, rooting in mud found in South and South-east Asia, Africa and Australia. The white globulous rhizomes which are harvested at 6–9 months measure 60–120 cm in length. The root yield is 5 t/ha and starch content is nearly 18%. In China, a fine starch is isolated from the rhizomes.

Oca is a small compact annual tuberous herb 20–30 cm high. Oca is a ancient food plant of the Andes and found in many parts of South America. The rhizomatous tubers are harvested at eight months maturity and are similar to potato (5–8 cm in length). The tuber yield is 4–5 t/ha while the starch content is 12%.

Queensland arrowroot is perennial herbaceous monocotyledon found in many parts of Asia, Africa and South America. The shape of the rhizomes varies from cylindrical to tapering and 5–9 cm in size. The tuber yield is 15–40 t/ha and the starch content varies from 24–30%.

Shoti known as Indian arrowroot, is a robust perennial with a fleshy branching rhizomes cultivated in Asia. The starchy finger shaped rhizomes are greyish in colour, grow to 15 cm in length and have a musky odour. The tuber yield is 8–12 t/ha and starch content is 12–15%. The starch is extracted by rasping the tubers, sieving and settling and serves as a source of easily digested starch.

Swamp taro is a giant herbaceous perennial 3–4 m in height with huge leaves. It is cultivated in many parts of Asia, Africa and the Pacific islands. The yield of corms is 7–10 t/ha and the starch content 28–30%.

Winged bean is a leguminous, climbing perennial found in Asia and Africa. Tubers are obtained 5–8 months after planting with a yield of 2–6 t/ha. Root tubers are 5–12 cm in length and contain nearly 20% starch.

Catch Crop

In agriculture, a catch crop is a fast-growing crop that is grown simultaneously with, or between successive plantings of a main crop. For example, radishes that mature from seed in 25–30 days can be grown between rows of most vegetables, and harvested long before the main crop matures. Or, a catch crop can be planted between the spring harvest and fall planting of some crops. Catch cropping is a type of succession planting. It makes more efficient use of growing space. Catch crops are also crops that are sown to prevent minerals being flushed away from the soil. By using catch crops, such as grain (millet) one can keep certain minerals not attached to the humous-clay connection (such as carbon (C) and other positively charged elements) in the soil for (many) years.

Here are some examples:

- Coriander or rocket sown midsummer after potatoes have been lifted. Both plants are quick to germinate and grow, before the autumn frosts. This also works even if you have grown your potatoes in a bag. Don't throw away the soil. Catch the last of the summer/autumn growing season by sowing some radishes or salad leaves.

- If you sow quick-growing radishes among rows of parsnips which are slow to germinate, the radishes will mark out the position of the parsnips before you see them. Thus you can weed around them without causing damage. And pull the radishes before the parsnips have formed.

- Lettuces sown beneath runner beans. You'll be cutting them before the beans are ready to pick. This is also known as inter-cropping.

The Best Catch Crops

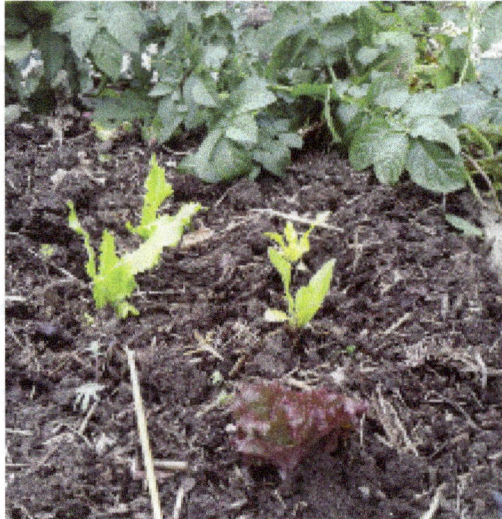

- Anything quick growing: salad leaves such as lettuce, mizuna and corn salad; radish; herbs such as coriander or rocket. Here you can see lettuces being grown as the potatoes are gradually lifted.

- Quick-to-grow small veg, including mini carrots, beetroots or turnips.

- Some green manures, such as clover. They feed the soil as they grow – often fixing nitrogen with their roots. If you have lifted your potatoes in early summer, sow the clover straight away. When you dig the foliage in, it will feed the soil, ready for winter or spring planting.

For catch crops to succeed, your soil must be in good heart. It needs to be full of rich nutrients to support this 'intensive' farming.

Permanent Crop

Permanent crops are perennial trees, bushes or vine crops like citrus, apples, blueberries, nuts or grapes. Although it represents a small segment in agriculture, some 11 percent (over 1.5 billion ha) of the globe's land surface (13.4 billion ha), it plays an important role in shaping the rural landscape and helping to balance agriculture within the environment.

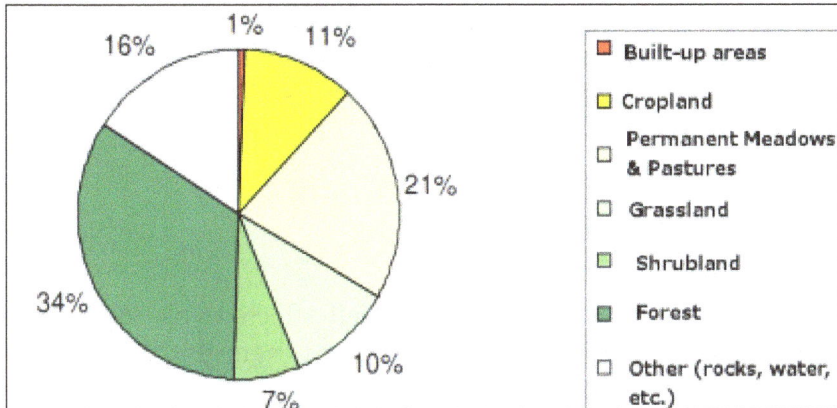

Different types of land utilisations (in % of land surfaces).

Because of high costs, planting permanent crops is considered to be a "high stakes" farming operation. Today, success is imperative. Permanent crop producers must minimize risk as much as possible to ensure successful crop establishment. As with most other high-risk farming operations, it is important to plan ahead. Planning ahead not only improves the chances of successful permanent crops establishment, but also greatly reduces the amount of personal worrying after it has been planted. Establishment depends on the complex interaction over planting time, soil, climatic, biotic, machinery and management factors. Natural conditions (climate, relief, soil) are fundamental factors that determine the choice for the establishment of permanent crops with economically justified production. According to Trent, certain fruit varieties can be grown in the following altitudes:

- Vineyard around 100-350 m,

- Peach and olive trees to 300 m,

- Chestnut to 600 m,

- Nut to 900 m,

- Pear and plum up to 1200 m,

- Apple and cherry to 1400 m,

- Hazel to 1800 m.

Increase in altitude changes temperature conditions. Thus, for every 100 m above sea level, median annual temperature decreases by 0.5-1 ° C, and the vegetation period by 7-10 days. It's best to raise wineyards and orchards on the hilly terrain to the south, southwest or southeast exposure that provide good insolation of orchard, and consequently a higher sugar content. Besides properly selected and prepared field for achieving high and quality yields, it is necessary to ensure optimum supply of all nutrient elements throughout the period of the growing season and also select rows direction. The most convenient row direction is the north-south. Rows should be placed in direction to rationalize all works: tillage, fertilization and protection. On sloping ground rows direction must be set so as to reduce erosion processes.

Crop Management

Crop management practices for the safe use of salt-affected soils and saline water primarily consists of growing suitable salt-tolerant crops, managing seedbeds, and grading fields to minimize local accumulation of salts, soil management, improving irrigation efficiency and soil, water, and salinity monitoring for assessing leaching and drainage requirements. Important crop management consideration includes:

- Appropriate crop genotypes for satisfactory yield under existing and predicted saline conditions.

- Appropriate planting procedures to minimize salinity near the vicinity of seed.

- Maintaining soil water at relatively high levels for plant growth.

- Appropriate soil management such as tillage, ploughing, sanding, and chemical amendments, mineral fertilizers, organic manures, and mulching to improve soil organic matter and increased soil percolation.

- An efficient water distribution system and irrigation scheduling.

Crop management begins with the sowing of seeds, continues with crop maintenance during growth and development, and ends with crop harvest, storage, and distribution. During seed sowing, a mechanized planter often opens a furrow in the prepared soil seed bed, places the seed in the exposed moist soil, covers the planted seed, and then often packs the soil down to assure firm seed-soil contact. In no-till systems, the crop is planted ("drilled") directly into the soil through residue from the previous crop.

Soil fertilization is an essential component of crop management to assure nutritional sufficiency for plant growth. The selection of type, amount, timing, and method of fertilizer application is determined by a variety of considerations including the crop type, the nature of the fertilizer, soil conditions, and weather. A generalized listing of common fertilizer applications follows:

- Broadcast (application of fertilizer (often pelletized) to the soil surface before the crop emerges);

- Plowing in (application of fertilizer to the surface followed by mixing into the topsoil by plowing);

- Sideband (fertilizer application in bands adjacent to the seed);

- Contact placement (fertilizer application in direct contact with the seed);

- Side-dressing (fertilizer placement in narrow rows at the surface after crop emergence);

- Top-dressing (general application of fertilizer to the crop after emergence).

During crop growth, a variety of crop and soil maintenance as well as weed removal practices may be undertaken. Again, the specific type of farm machinery actually used and overall management practices are site, farmer, and climate specific. But overall, weed control can be accomplished through several types of soil cultivation practices. These include dense arrays of small spring tines, rotary hoes, and tractor-mounted arrays of spear- or sweep-pointed shanks designed to till in between crop rows. Herbicides are also used widely for weed control.

In essence, crop management practices influence the subsurface habitat by two independent mechanisms. First, the physical structure of soil (hence the infiltration rates of water) is altered by farm machinery traffic passing over the soil, by cultivation implements, and by the penetration of soil by roots and shoots of the growing crop plants. Second, the solutes in soil that may be conveyed to the subsurface by infiltrating water are determined by the organic and inorganic compounds present in the soil as a result of fertilizer amendments and crop growth and decay.

Crop management can influence the nature of thrips injury in several ways. For example, the proximity of susceptible crops to thrips sources is important. Damage sometimes occurs when thrips disperse in large numbers into susceptible crops. This often results when an early season crop such as oats or wheat reaches maturity, or when a crop is cut at mid-season, as is the case with alfalfa and clover.

Mulches can influence the abundance of thrips and the transmission of plant viruses. In studies conducted in Louisiana, aluminum-surfaced mulch reduced the incidence of tomato-spotted wilt transmission by thrips to tomato and pepper by about 60–80%. This approach toward disease management has been studied much more with aphid vectors, but most of this technology is probably applicable to thrips-transmitted diseases.

Intercropping can have some benefit for onion thrips management. Despite its wide host range, there are clearly preferred hosts, principally onion. For example, Uvah and Coaker alternated rows of onion with various ratios of carrot rows, and found that the presence of carrots decreased abundance of thrips. This occurred despite the fact that carrot is a nominal host of thrips.

Sanitation is very important. Long ago, Horsfall and Fenton noted the ability of thrips to disperse from contaminated overwintering plants left in the field, or from transplanted onion bulbs taken from storage, to newly seeded onions. Now, with the availabilty of rapid transportation, thrips are often moved with plant material, and then inadvertently inoculated into fields. For example, Schwartz et al. found that nearly all batches of onion transplants shipped from Texas to Colorado were contaminated with thrips. Sporadic incidence of insecticide resistance among Colorado onion fields apparently was related to different sources of onions and thrips, and different pesticide exposure histories. Also, in some northern areas greenhouses are a source of thrips in the spring.

Pollination Management

Pollination management is the label for horticultural practices that accomplish or enhance pollination of a crop, to improve yield or quality, by understanding of the particular crop's pollination needs, and by knowledgeable management of pollinizers, pollinators, and pollination conditions.

While people think first of the European honey bee when pollination comes up, in fact there are many different means of pollination management that are used, both other insects and other mechanisms. There are other insects commercially available that are more efficient, like the blue orchard bee for fruit and nut trees, local bumblebees better specialized for some other crops, hand pollination that is essential for production of hybrid seeds and some greenhouse situations, and even pollination machines.

Pollinator Decline

With the decline of both wild and domestic pollinator populations, pollination management is becoming an increasingly important part of horticulture. Factors that cause the loss of pollinators include pesticide misuse, unprofitability of beekeeping for honey, rapid transfer of pests and diseases to new areas of the globe, urban/suburban development, changing crop patterns, clear cut logging (particularly when mixed forests are replaced by monoculture pine), clearing of hedgerows and other wild areas, bad diet because of loss of floral biodiversity, and a loss of nectar corridors for migratory pollinators.

Importance

The increasing size of fields and orchards (monoculture) increase the importance of pollination management. Monoculture can cause a brief period when pollinators have more food resources than they can use (but monofloral diet can reduce their immune system) while other periods of the year can bring starvation or pesticide contamination of food sources. Most nectar source and pollen source throughout the growing season to build up their numbers.

Crops that traditionally have had managed pollination include apple, almonds, pears, some plum and cherry varieties, blueberries, cranberries, cucumbers, cantaloupe, watermelon, alfalfa seeds, onion seeds, and many others. Some crops that have traditionally depended entirely on chance pollination by wild pollinators need pollination management nowadays to make a profitable crop. Many of these were at one time universally turning to honeybees, but as science has shown that honeybees are actually inefficient pollinators, demand for other managed pollinators has risen. While honeybees may visit dozens of different kinds of flowers, diluting the orchard pollen they carry, the Blue orchard bee will visit only the intended tree, producing a much higher fertilization rate. The focus on the specific tree also makes the orchard bee 100 times more efficient at pollinating, per bee.

Some crops, especially when planted in a monoculture situation, require a very high level of pollinators to produce economically viable crops, especially if depending on the more generalized honeybee. This may be because of lack of attractiveness of the blossoms, or from trying to pollinate with an alternative when the native pollinator is extinct or rare. These include crops such as alfalfa, cranberries, and kiwifruit. This technique is known as saturation pollination. In many such cases, various native bees are vastly more efficient at pollination (e.g., with blueberries), but the inefficiency of the honey bees is compensated for by using large numbers of hives, the total number of foragers thereby far exceeding the local abundance of native pollinators. In a very few cases, it has been possible to develop commercially viable pollination techniques that use the more efficient pollinators, rather than continued reliance on honey bees, as in the management of the alfalfa leafcutter bee.

In the case of the kiwifruit, its flowers do not even produce nectar, so that honeybees are reluctant to even visit them, unless present in such overwhelming numbers that they do so incidentally. This has led bumblebee pollination companies to begin offering their services for kiwifruit, as they appear to be far more efficient at the job than honeybees, even more efficient than hand pollination.

Number of hives needed per unit area of crop pollination		
Common name	Number of hives per acre	Number of hives per hectare
Alfalfa	1, (3–5)	2.5, (4.9–12)
Almonds	2–3	4.9–7.4
Apples (normal size)	1	2.5
Apples (semi dwarf)	2	4.9
Apples (dwarf)	3	7.4
Apricots	1	2.5
Blueberries	3–4	7.4–9.9
Borage	0.6–1.0	1.5–2.5
Buckwheat	0.5–1	1.2–2.5

Canola	1	2.5
Canola (hybrid)	2.0–2.5	4.9–6.2
Cantaloupes	2–4, (average 2.4)	4.9–9.9, (average 5.9)
Clovers	1–2	2.5–4.9
Cranberries	3	7.4
Cucumbers	1–2, (average 2.1)	2.5–4.9, (average 5.2)
Ginseng	1	2.5
Muskmelon	1–3	2.5–7.4
Nectarines	1	2.5
Peaches	1	2.5
Pears	1	2.5
Plums	1	2.5
Pumpkins	1	2.5
Raspberries	0.7–1.3	1.7–3.2
Squash	1–3	2.5–7.4
Strawberries	1–3.5	2.5–8.6
Sunflower	1	2.5
Trefoil	0.6–1.5	1.5–3.7
Watermelon	1–3, (average 1.3)	2.5–4.9, (average 3.2)
Zucchini	1	2.5

It is estimated that about one hive per acre will sufficiently pollinate watermelons. In the 1950s when the woods were full of wild bee trees, and beehives were normally kept on most South Carolina farms, a farmer who grew ten acres (4 ha) of watermelons would be a large grower and probably had all the pollination needed. But today's grower may grow 200 acres (80 ha), and, if lucky, there might be one bee tree left within range. The only option in the current economy is to bring beehives to the field during blossom time.

Types of Pollinators

Organisms that are currently being used as pollinators in managed pollination are honey bees, bumblebees, alfalfa leafcutter bees, and orchard mason bees. Other species are expected to be added to this list as this field develops. Humans also can be pollinators, as the gardener who hand pollinates her squash blossoms, or the Middle Eastern farmer, who climbs his date palms to pollinate them.

The Cooperative extension service recommends one honey bee hive per acre (2.5 hives per hectare) for standard watermelon varieties to meet this crop's pollination needs. In the past, when fields were small, pollination was accomplished by a mix of bees kept on farms, bumblebees, carpenter bees, feral honey bees in hollow trees and other insects. Today, with melons planted in large tracts, the grower may no longer have hives on the farm; he may have poisoned many of the pollinators by spraying blooming cotton; he may have logged off the woods, removing hollow trees that provided homes for bees, and pushed out the hedgerows that were home for solitary native bees and other pollinating insects.

Planning for Improved Pollination

US migratory commercial beekeeper moving spring bees from South
Carolina to Maine for blueberry pollination.

Before pollination needs were understood, orchardists often planted entire blocks of apples of a single variety. Because apples are self-sterile, and different members of a single variety are genetic clones (equivalent to a single plant), this is not a good idea. Growers now supply pollinizers, by planting crab apples interspersed in the rows, or by grafting crab apple limbs on some trees. Pollinizers can also be supplied by putting drum bouquets of crab apples or a compatible apple variety in the orchard blocks.

The field of pollination management cannot be placed wholly within any other field, because it bridges several fields. It draws from horticulture, apiculture, zoology (especially entomology), ecology, and botany.

Improving Pollination with Suboptimal Bee Densities

Growers' demand for beehives far exceeds the available supply. The number of managed beehives in the US has steadily declined from close to 6 million after WWII, to less than 2.5 million today. In contrast, the area dedicated to growing bee-pollinated crops has grown over 300% in the same time period. To make matters worse, in the past five years we have seen a decline in winter managed beehives, which has reached an unprecedented rate near 30%. At present, there is an enormous demand for beehive rentals that cannot always be met. There is a clear need across the agricultural industry for a management tool to draw pollinators into cultivations and encourage them to preferentially visit and pollinate the flowering crop. By attracting pollinators like honeybees and increasing their foraging behavior, particularly in the center of large plots, we can increase grower returns and optimize yield from their plantings.

References

- Crop-agriculture, topic: britannica.com, Retrieved 14 February, 2019

- manjul, tarannum (january 21, 2006). "millets older than wheat, rice: archaeologists". Lucknow newsline. Archived from the original on may 9, 2008. Retrieved 2008-04-14

- Classification-of-crops-7-categories-crop-production, production-crops, crops: yourarticlelibrary.com, Retrieved 10 May, 2019

- D'andrea, a. C.; casey, j. (2002). "pearl millet and kintampo subsistence". The african archaeological review. 19 (3): 147–173. Doi:10.1023/a:1016518919072. Issn 0263-0338. Jstor 25130746

- Tuber-crops, agricultural-and-biological-sciences, topics: sciencedirect.com, Retrieved 29 April, 2019

- Lawler, a. (2009). "bridging east and west: millet on the move". Science. 325 (5943): 942–943. Doi:10.1126/science.325_940. Pmid 19696328

- Cereal-crop, earth-and-planetary-sciences, topics: sciencedirect.com, Retrieved 19 August, 2019

- Berry, wendell (5 january 2009). "a 50-year farm bill". A 50-year farm bill. The new york times. Retrieved 25 march 2011

- Permanent-crop-establishment, post: agrivi.com, Retrieved 1 July, 2019

- Catch-crop, dic-en: educalingo.com, Retrieved 25 May, 2019

- Borders, max; burnett, h. Sterling (march 24, 2006). "farm subsidies: devastating the world's poor and the environment". National center for policy analysis. Retrieved april 6, 2012

- Forage-crops, grazing-pasture-management, research-and-development: mla.com.au, Retrieved 30 June, 2019

- Crop-management, agricultural-and-biological-sciences, topics: sciencedirect.com, Retrieved 1 January, 2019

- Frank j. Dainello & roland roberts. "cultural practices". Texas vegetable grower's handbook. Texas agricultural extension service. Retrieved 9 december 2014

2

Types of Cropping

Multicropping, intercropping and monocropping are the main types of cropping. Multicropping includes the production of more than one crop in the same piece of land in the same seasons and intercropping is the practice of growing different kinds of crop in close proximity. The topics elaborated in this chapter will help in gaining a better perspective about these various types of cropping.

Multicropping

Multicropping is not a new form of agricultural technology, but instead is an ancient means of intensive farming. Multicropping has been practiced in many parts of the world as a way to maximize land productivity in a specific area in a growing season. Generally, the practice of planting two or more crops on the same field is more common in tropical regions where more rainfall, higher temperatures, and longer growing seasons are more favorable for continual crop production. As population has increased, increasing the need for agricultural production, the use of multicropping systems is more prevalent. Though the history of Multicropping is old, the concept has received very little attention from agricultural scientists, and what limited interest exists has come about very recently.

Why was this interest increased so dramatically in such a short time? Food shortages in many parts of the world, as well as the threat of insufficient supplies in the near future, continues to stimulate more intensive agricultural investigation in a search for more productive alternatives. As a consequence, it appears that we are about to embark on a new phase of agricultural research. Exactly what form it will take is still not known, but the reasons for this new approach are rapidly becoming apparent.

First, we have begun to observe a leveling off in yield increases brought about by the types of genetic manipulation that gave us such rapid and impressive yield increases during the "Green Revolution." It is as if we have reached a "yield plateau" with the current lines of research and crop selections. Large-scale use of single varieties (e.g., some of the International Rice Research Institute (IRRI) varieties of rice), with broad adaptability, produced major breakthroughs in yields. But it appears that these varieties have almost reached their maximum yield potentials. In many areas with specific soil and climatic conditions, they have not performed as well as hoped, especially on land more difficult to mechanize or irrigate. Thus we must begin to look for varieties with more specific adaptability and selected for specific environments, or else consider alternative cropping systems.

Second, most of the dramatic yield increases during the past few decades have been on the best agricultural lands–areas with good soil and easy water control. Future increases in production, therefore, will demand a new and innovative way of managing these highly productive lands, as well as looking for methods to make marginal lands increasingly productive. Only 20 percent of Asia rice land, for example, is irrigated, and the new high yielding rice varieties (which also require high levels of fertilizers, water use, and pest control) have not penetrated much beyond this boundary.

The third factor is the oil crisis. Oil prices continue to soar, and with them, the cost of fertilizers, pesticides, and fuel needed to build and run farm equipment and move irrigation water. Costs continue to mount for those inputs most responsible for achieving the dramatic yield increases of the "Green Revolution." We are faced with the necessity of having to consider other alternatives that might allow us to substitute innovative biological or agronomic practices and varieties for these high cost inputs. Multicropping offers one of the most important and promising of these alternatives.

Multicropping systems use management practices where the total crop production from a single piece of land is achieved by growing single crops in close sequence, growing several crops simultaneously, or combining single and mixed crops in some sequence. The most important aspect of multicropping is the intensification of crop production into additional dimensions. Multicropping includes the dimensions of time and space; for example, when two crops share the same space at the same time.

A classification of types of multicropping systems is presented in table. Note that special emphasis is placed on the distinction between intercropping, where two or more crops are grown at the same time, and sequential cropping, where two or more crops are grown on the same piece of land, but one following the other.

Agroforestry, as a particular type of intercropping system, will be discussed in some detail. Also, "mixed cropping", "polyculture," and "multicropping" will be used interchangeably in this review. By combining different aspects of simultaneous and sequential cropping systems, it is possible to visualize a truly complex pattern of different multicropping systems. This classification will be used throughout the following discussion, based on a symposium sponsored by the American Society of Agronomy, in support of the need to standardize terminology.

Principal Multicropping Patterns

- Multicropping: The intensification of cropping in time and space dimensions. Growing two or more crops on the same field in a year.

- Intercropping: Growing two or more crops simultaneously on the same field per year. Crop intensification is in both time and space dimensions. There is inter-crop competition during all or part of crop growth. Farmers manage more than one crop at a time in the same field.

  - Mixed intercropping: Growing two or more crops simultaneously with no distinct row arrangement.

  - Row intercropping: Growing two or more crops simultaneously with one or more crops planted in rows.

  - Strip intercropping: Growing two or more crops simultaneously in different strips wide enough to permit independent cultivation but narrow enough for the crops to interact agronomically.

  - Relay intercropping: Growing two or more crops simultaneously during part of each one's life cycle. A second crop is planted after the first crop has reached its reproductive stage of growth, but before it is ready for harvest.

- Sequential Cropping: Growing two or more crops i n sequence on the same field per year. The succeeding crop is planted after the preceding one has been harvested. Crop intensification is only in the time dimension. There is no intercrop competition. Farmers manage only one crop at a time.

  - Double cropping: Growing two crops a year in sequence.

  - Triple cropping: Growing three crops a year in sequence.

  - Quadruple cropping: Growing four crops a year in sequence.

  - Ratoon cropping: Cultivating crop regrowth after harvest, although not necessarily for grain.

The Basis of Multicropping

Yield Advantages of Crop Mixtures

In areas of the world where multicropping is a common aspect of agroecosystem management, productivity generally is more stable and constant in the long term. Farmers often are able to achieve a combined production per unit area greater with a crop mixture than with an equal area divided among separate crop units. In such cases the Relative Yield Total (RYT) is greater than 1.0. It may be that each crop in the mixture yields slightly less than the monoculture, but the combined yield of the mixture on less total land area is the important aspect.

In one study, the results of 572 comparisons of crop mixtures demonstrated that the majority (66 percent) had RYTs close to 1.0, indicating no distinct advantage to the mixture. On the other hand, 20 percent of the mixtures had RYTs greater than 1.0, ranging up to 1.7, indicating advantages to the mixtures. Only 14 percent had less than 1.0, indicating distinct disadvantages. It must be remembered that most of the cases studied were experimental planting and not actual multicropping systems. Farmers would tend to choose the systems that yield more, as we have observed in traditional agroecosystems in the lowland tropical areas of southeastern Mexico.

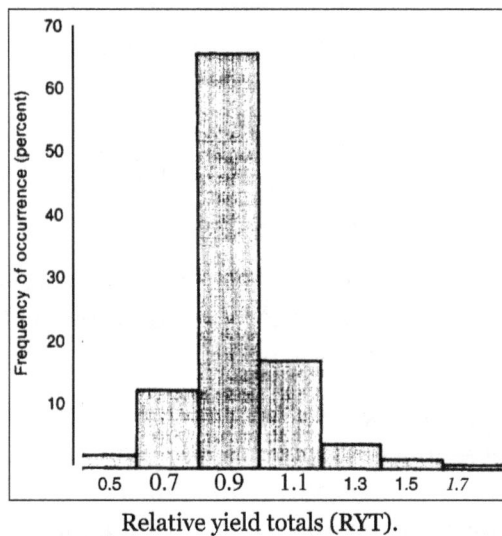

Relative yield totals (RYT).

The fact that advantageous mixtures do exist demonstrate the need for detailed research to take proper advantage of such systems. But for such systems to be considered as actual alternatives we need to understand thoroughly the biological and agronomic basis responsible for the observed response, as well as the advantages and disadvantages to their use. Before beginning a discussion of each aspect, a basic outline of such characteristics is presented, separated broadly into biological and physical aspects and socioeconomic aspects. In many cases it is understood that there may be overlap between the two classifications, yet it is hoped that in the course of the following discussion that such aspects will be clarified.

Advantages

- It is possible to obtain a better use of vertical space and time, imitating natural ecological patterns in regards to structure of the system, and permitting efficient capture of solar energy and nutrients.

- Greater amounts of biomass (organic matter) can be returned to the system, sometimes even of better quality.

- There exists a more efficient circulation of nutrients, including their "pumping" from the deeper soil profiles when deeper rooted shrubs or trees are included.

- The damaging effects of wind sometimes can be reduced.

- Systems can be designed that are appropriate for (but not restricted to) marginal areas because multicropping systems can better take advantage of variable soil, topography, and steeper slopes.

- Multicropping systems are less subject to variability in climatic conditions, especially extremes of rainfall, temperature, or wind.

- Reduction of water evaporation from the soil surface.

- Increased microbial activity in the soil.

- Avoidance or reduction of surface erosion.

- Fertilizer use can be more efficient because of the more diverse and deeper root structure in the system.

- Improved soil structure, avoiding the formation of a "hard pan" and promoting better aeration and filtration.

- Legumes (as well as a few other plant families) are able to fix and incorporate nitrogen into the system.

- Heavier mulch cover aids in weed control.

- Better opportunities for biological control of insects and diseases.

- Crop mixtures better permit the functioning of complex mutualisms and beneficial interactions between organisms.

- Better use of time, with more crops per unit time in the same area.

- Dependence on one crop is avoided so that variability in prices, market, climate, and pests and diseases do not have such drastic effects on local economics.

- Less need to import energy, pay for fertilizers, pay for externally produced materials, or depend on machinery.

- Wildlife is favored, and with rational use it can be an important source of protein.

- Greater flexibility of the distribution of labor over the year.

- Recovery of investments can occur in much less time, especially where trees are combined with short term agricultural crops.

- Harvest is spread over a longer period of time.

- In areas and times of high unemployment, multicropping systems can use much more labor.

- Farmers can produce a large variety of useful products, depending on the type and complexity of the multicropping systems, such as firewood, construction materials, flowers, honey, crops for home consumption, thus lowering the out-flow of funds.

- Certain multicropping systems permit a gradual change from destructive farming practices to more appropriate technologies, without a drop in productivity.

- Multicropping can promote a return to the land, and its maintenance.

- In systems which include trees and/or animals, such components can constitute a type of "savings" for the future, while short term crops satisfy immediate needs.

- Because of their diverse nature, multicropping systems promote interdisciplinary activities, stimulate interchange and group activities, and lead to social cohesion in the long term.

Disadvantages

- Competition between plants for light.

- Competition between plants for soil nutrients.

- Competition between plants for water.

- Possibility for allelopathic influences between different crop plants due to plant-produced toxins. Harvesting of one crop component may cause damage to the others.

- It is very difficult to incorporate a fallow period into multicropping systems, especially when long lived tree species are included.

- It is sometimes impossible, and many times very difficult, to mechanize multiple crop systems. Increased evapotranspiration loss of water from the soil, caused by greater root volume and larger leaf surface area.

- Possible over-extraction of nutrients, followed by their subsequent loss from the system with the increased exportation of agricultural or forest products.

- Leaf, branch, fruit, or water-drop fall from taller elements in a mixed crop system can damage shorter ones.

- Higher relative humidity in the air can favor disease outbreak, especially of fungi.

- Possible proliferation of harmful animals (especially rodents and insects).

- The systems are more complex and less understood agronomically and biologically. Statistical designs for experimental analysis are much more complex.

- Yields sometimes are lower, providing only subsistence level production.

- In many systems, multicropping is not considered to be economically efficient due to the complexity of activities necessary.

- These systems require more hand labor, which can be considered a disadvantage in some circumstances.

- Some mixed crop systems do not offer sufficient reward to lower income farmers to raise their standard of living.

- For producers with limited economic resources, it may take longer to recover the entire initial investment.

- Farmers initiating multicropping systems may encounter opposition from the prevalent social, economic, and political system.

- There is a shortage of trained personnel (technical and scientific) capable of installing and managing multicropping systems.

- There is a general lack of knowledge or understanding of multicropping by "decision makers," affecting especially funding for research to make such systems viable alternatives.

The most commonly accepted reason explaining why it is possible to obtain better yields with crop mixtures is that the component crops differ in their growth requirements. Such combinations of components can be said to be "complementary".

A mixture makes better overall use of available resources. Negative influences between the component members of a successful multicropping system would be reduced considerably. To maximize the advantages of such a system, it is important to maximize the degree to which one component complements another. With a greater range of requirements between different elements of the mixture, theoretically the greatest advantages would be achieved.

One way to achieve complementarily is by varying the crop components temporally—using sequential planting to achieve a multicropping system that ensures that antagonistic interactions between the components are avoided. Following a crop with another that has different growth requirements would enable the maximum use of resources. This concept has been used for a long time and is the basic rationale behind crop rotations.

The most advantageous use of soil, for example, would be to follow one crop with another that requires different soil nutrients. A subsequent crop would thus be able to absorb fertilizer residues left over from the previous crop, thus reducing the need for fertilizer applications. For the Eastern United States, it has been concluded that double cropping systems such as soybeans after wheat or barley, or the production of silage crops after grain corn or sorghum, can function well, depending on the length of the growing season, numerous sequential plantings can take place during a single year. Such systems require special management, with timely harvest, use of proper varieties, alteration of the standard planting distance, special selection of herbicides so as to not create antagonisms or residual effects, and also the possibility of using no-tillage planting with certain of the row crops.

Another form of complementing different crop components is through an intensification of the sequential cropping system known as relay planting. The same avoidance of overlapping plant growth requirements is gained, as well as the avoidance of direct plant interference, by planting a second crop after the first one has completed the major part of its development, but before harvest. Relatively little research on relay cropping has been done in the United States, and most has demonstrated little if any yield advantage. On the other hand, in Mexico and Latin America innumerable examples of relay planting with definite yield advantages have been reported, especially for corn and beans.

Again, the important, and as yet little studied, aspect of relay planting success depends on the correct combinations of timing and varieties so as to avoid shading, nutrient competition, or inhibition brought about by toxicity produced by the decomposition of a previous crop residue.

Finally, maximum complementarily can be achieved by growing two or more crops simultaneously, either in rows, strips, or mixed, but taking advantage of the spatial arrangement of the different crops and knowledge of their individual growth requirements. Again, most examples of such systems come from outside the United States. One particularly well-documented example is a traditional corn, bean, and squash system in Tabasco, Mexico.

Corn is planted at a density of 50,000 plants/ ha, climbing beans in the same hole at a density of 40,000 plants/ha, and the squash intermixed among the rows of corn and beans at a density of 3,330 plants/ha. All are planted at the same time in this case. Beans begin to mature first, using the corn stalks for support; the corn matures second; the squash is the last to mature. Aerial space is divided such that corn occupies the upper canopy, beans the middle, and squash covers the ground. Better weed control is achieved, and insect pests are largely controlled by natural enemies. Corn yield was significantly higher for the polyculture as compared to different densities of monoculture, but beans and squash suffered a distinct yield reduction. Interestingly, the LER (Land Equipment Ratio) value of 1.73 tells us that the sum of the yields in the mixture can only be equaled in monoculture by planting 1.73 times the area divided proportionally among the three sole crops.

Table: Yields of Corn, Beans, and Squash (kg/ha) Planted in Polyculture as Compared to Low and High Densities of Each Crop in Monoculture.

	Total grain or fruit yields				
Crop	Monoculture				Polyculture
Corn:					
Density ... ,	33,300	40,000	66,600	100,000	50,000
Yield	990	1,150	1,230	1,170	1,720
Beans:					
Density	56,800	64,000	100,000	133,200	40,000
Yield	425	740	610	695	110
Squash:					
Density	1,200	1,875	7,500	30,000	3,330
Yield	15	250	430	225	80

Crop	Total biomass dry weight				
Corn	2,822.9	3,119.4	4,477,5	4,870.9	5,927.2
Beans	852.9	895.1	842.6	1,390.4	253.1
Squash	240.9	940.9	1,254.0	801.9	478.3
		Total Polyculture Biomass			6,658.6

LER (Land Equipment Ratio) = $\dfrac{\text{Sum } of\ yields\ of\ each\ polyculture}{\text{Sum of highest yield each monoculture}}$

$$LER = \frac{1,720}{1,230} + \frac{110}{740} + \frac{80}{430}$$

$$LER = 1.73$$

The advantage of producing a greater yield altogether on less land is obvious. The much higher total yield of biomass in the mixture is also important because much of this organic matter is returned to the soil, bringing important consequences in soil fertility, humidity conservation, microbial activity, etc., all related to the success of the following crops, Currently, studies are being conducted to determine if the higher yields are the result of more efficient resource use, or if in fact some mutually beneficial effect between crop components is taking place, for example, the bean producing nitrogen that the corn can absorb. This example demonstrates the enormous potential that multicropping systems offer for the future.

Specific Resource use, Conservation and Management

An intensified land-use system of agriculture will certainly put greater pressures on the available natural resources of our crop and rangelands. Considerable discussion has focused on the harmful or beneficial aspects of this intensification, and a review of some of the more important aspects can aid greatly in understanding this problem:

1. Microclimate and Light: In any agroecosystem, a very important aspect of productivity is related to the amount of light converted directly to carbohydrate, hence to vegetative

material, through photosynthesis. Each cropping system has a photosynthetic potential, based on its capacity of conversion. Monoculture, especially of annuals, generally
have a lower potential because either the plant cover is not complete, or the soil is occupied only during one short season, leaving the surface bare of photosynthetic capacity
until the next crop is planted. Light is not like other resources, where a reservoir exists
and the plants tap it as the need arises. Rather, it has to be used when it is available,
thus leaf area becomes a very important factor. A multi-layered polyculture would be
able to capture much more light energy, raising efficiency, and potentially, production.

Apart from the quantity of light absorbed, its quality is also important. Light that has
passed through a leaf layer is altered as certain light waves are absorbed and others penetrate. Plants in the lower layers of the canopy need to be adapted to this alteration—an
aspect well studied only in natural vegetation. For crop ping systems, light has been
studied in detail only for monoculture systems from the point of view of increasing
effective photosynthetic leaf area for the single crop. By manipulating species with
different light requirements, greater photosynthetic potential can be achieved. This is
made easier by using dominant species in the polyculture that do not develop a closed
canopy, allowing considerable penetration to the next levels. The most shade-tolerant
plants should be in the lowest levels. In such a system, the soil surface is in essence
completely covered by plants. This manipulation of plant architecture has been studied
in detail ecologically and has considerable application in multicropping systems.

Other aspects of the crop microclimate are also affected. Crops in the lower layers would
be subject to less water stress, but care must be taken that root system competition for
water does not become a problem. Water loss by soil surface evaporation could be reduced, but transpiration from leaf surfaces might be increased in the crop mixture. Soil
temperatures would be lowered, an advantage especially in warmer and drier environments, aiding in the conservation and buildup of organic matter in the soil. Protection
from wind would be provided for the lower canopy species. Care would need to be taken
that the increased humidity in the lower canopies does not promote higher incidence of
certain diseases, especially fungi, either of the roots or foliage.

2. Soil-Plant Relations in Multicropping Systems: Any time that we try to combine two
or more crops simultaneously in one area, there exists the possibility for complex interactions between the plants and their soil environment. When total complementarily is
achieved, the roots of the component species occupy different soil horizons, reducing
considerably the potential competition between species and increasing the efficiency of
total nutrient uptake. In combinations of deep rooted with shallow-rooted species, especially when trees are planted with grasses or annual crops, the trees are capable of
absorbing uncaptured nutrients as they are leached into the soil. Then, through their
transport to foliage, they can be deposited on the soil surface again as the leaves drop.

Intercropping systems have been shown to extract more nutrients from the soil than do
single crop plantings per unit area of land. In a very complete study with corn and pigeon

peas in Trinidad, various parameters of crop response were measured. The highest single crop yields of grain were obtained in monoculture, but by adding yields of two crops planted mixed or in intercropped rows, Relative Yield Totals (RYT) were higher. Total dry matter production was higher in the mixtures as well. The most interesting aspect is the uptake of nutrients (N, P, K, Ca, and Mg). The total uptake is based on the sum of the two crops together, and in all cases the total nutrient content of the dry matter production was higher for the mixtures, demonstrating the greater extractive capacity of the multicropping system. Apparently, for corn and pigeon peas, row intercropping gave the best results, demonstrating that at times two crops together can negatively influence each other, but the total yield makes up for the reduction. Each crop mixture needs to be examined in detail.

The greater uptake of nutrients in crop mixtures could deplete the soil more rapidly. But an aspect of multicropping that needs to be considered is what proportion of this nutrient content is removed from the system with the harvest, as compared to the part reincorporated back into the system. In table, a corn/bean polyculture is compared to a corn monoculture. Total biomass production, as well as yield removed from the system, is considerably higher from the mixture (10,24 tons/ ha versus 6.68 tons/ha total biomass). The percentage of this total that leaves the system is slightly lower for the mixture (61 percent versus 66 percent), but the actual amount of organic matter returned to the soil in the polyculture (3.98 tons/ha) as compared to the sole crop (2.32 tons/ ha) demonstrates that although more material is produced by the intercrop system, a greater amount returns to this system. This possibly offsets any increase in extraction of soil nutrients and permits the long-term management of the system.

Table: Effects of Mixed and Row Intercropping on Yields and Nutrient Uptake of Corn (C) and Pigeon Peas (PP) in St. Augustine, Trinidad, Expressed as Relative Yield Totals (RYT).

Parameter	Sole crop		Mixed intercrop			Row intercrop		
	C	PP	C	PP	RYT	C	PP	RYT
Grain yields (tons/ha)	3.1	1.9	2.0	1.7	1.54	2.6	1.8	1,78
Total Dry Matter (tons/ha)	6.4	5.1	4.2	3.8	1.40	5.0	4.9	1.74
N uptake (kg/ha)	-66.0	119.0	48.0	100.0	1.56	54.0	127.0	1.88
P uptake (kg/ha)	13.0	6.0	9.0	5.0	1.52	11.0	7.0	2.01
K uptake (kg/ha)	51.0	37.0	37.0	32.0	1.59	46.0	33.0	1.79
Ca uptake (kg/ha)	10.0	22.0	10.0	15.0	1.68	9.0	19.0	1.76
Mg uptake (kg/ha)	12.0	14.0	9.0	8.0	1.32	9.0	12.0	1.61

Table: Biomass Distribution (in tons/ha) of Dry Matter in a Corn/Bean Polyculture as Compared to a Corn Monoculture, in Tacotalpa, Tabasco, Mexico.

Crop	Roots	Crown	Leaves and stem	Grain[a]	(A) Total	(B) Removed matter	(B) (A) percent	(A)-(B) Total reincorporated
Corn	0.49	0.60	2.29	4.76[b]				
plus					10.24	6.26	61 %	3.98
Beans	0.15	0.00	0.45	1.50[b]				
Corn								
Alone	0.34	0.41	1.57	4.36[b]	6.68	4.36	65 %	2.32

Weight of grain of corn is unhusked, including cob and husk, in the manner that the harvest is removed from the field in this region. bIndlcates the removed portion of the biomass.

Another way to increase the return of nutrients to the system is to plant "nurse plants." These plants do not contribute directly to the biomass harvested and removed from the system, but their capacity to capture nutrients and continually recycle them in the soil would be an advantage. Local farmers in Tabasco, Mexico, use this concept in the management of weeds, leaving those that don't interfere with the crops and removing those that are harmful. This practice also provides a constant cover over the soil and helps maintain better soil structure, conserves water, fosters more microbial activity, and over the long run, requires fewer chemical fertilizers. By including plants that "trap" nutrients, such as legumes, such benefits can be improved even more. The widespread use of legume trees for shade in coffee and cocoa plantations is a classic example.

3. Water Use in Multicropping Systems: Any discussion of water use should consider rooting patterns. In multicropping systems, especially with several crops with differently arrayed root systems, a greater volume of the soil typically is occupied and thus water use efficiency is higher. This is useful, on the one hand, in areas where water supplies are limited. It also helps make more complete use of costly irrigation water. It has been proposed that cover crops in orchards stimulate deeper rooting by the trees. Different peak periods of water use in the crop mixtures would avoid competition and increase overall water use efficiency. A crop such as corn that uses relatively little water in its early stages of development could be interplanted with an early maturing crop that could take advantage of the unused moisture.

In areas where water is severely limited, care must be taken not to plant crops with overlapping water requirements because in dry years one member of the mixture could be outcompeted by the other. Combining two crops with slightly overlapping water needs, on the other hand, could be used to an advantage in areas with widely fluctuating rainfall regimes. In a dry year, one component would be favored, and in a wet year the other, guaranteeing profitable harvests of at least one crop every year. Studies on water availability in each region, coupled with studies of water needs of each component crop of multicropping systems, are critical for proper management.

The important effects of multicropping on the conservation of water and soil are primarily achieved through the maintenance of a more complete vegetative cover over the soil. It is important to remember that apart from improving cover while the crop is growing, multicropping systems aim toward maintaining this cover between harvests. This is achieved by reducing the time between harvest and replanting in sequential systems, planting a new crop into another in relay cropping, and continually interplanting in an intercropped system. The use of trees, either as windbreaks, for soil stabilization on eroded hillsides, or in areas subject to desertification, can be enhanced greatly by combining them with crops or pasture grasses.

Although it appears that multicropping systems use more water, their ability to obtain water not available to monoculture, use the water more efficiently, and contribute significantly to soil conservation, demonstrate a further potential for their more widespread use.

4. Pest, Disease, and Weed Relations: As discussed, possibilities exist for multicropping systems to be both advantageous and disadvantageous in relation to problems of pests, diseases, and weeds. The problem has to do with the great complexity of environmental factors and their dynamic interactions within the cropping systems. Where capital is not available or technical assistance has not been accepted, we observe that the main means of pest, disease, and weed control is through biological control, and through the management of a great diversity of cropping patterns, both in time and space.

It has been suggested that multicropping systems permit such a control because they are much less subject to attack. This comes about because the mixed cropping system prevents spread of diseases and pests by separating susceptible plants one species sometimes serves as a trap crop, protecting the others associated species sometimes serve as a repellant of the pest or disease to which the other crops are subject a greater abundance of natural predators or parasites of pests are present due to a higher diversity of adequate microsites and alternate prey.

However, there are also reasons why a multicropping system may be more susceptible to attack: reduced cultivation and greater shading due to the presence of associated species associated crops serve as alternate hosts crop residues from one crop may serve as a source of inoculum for the others. All of these advantages and disadvantages can exist, and further study is necessary to achieve the combinations that give the most positive results.

A few examples might serve to demonstrate the potential of multicropping for biological control. In one study, it was shown that the planting of a locally used medicinal herb (Chenodium ambrosioides) in sequence with corn or beans reduced the incidence of nematode populations in the soil, demonstrating a potential for reducing attack on the roots of the food crops. The herb added substances toxic to the nematodes into the soil. In another study, yields of cotton untreated with insecticides, but interplanted with sorghum, were 24 percent higher than sprayed monoculture. The reason was that sorghum served as a microhabitat for cotton bollworm predators. In another case, fall army worms were less a problem on corn associated with bush beans than on pure-stand corn. Beans intercropped with corn were attacked less by rust compared to beans in pure stands, probably because corn functions as a barrier to the dissemination of the fungal spores.

Weeds, on the other hand, present another problem. It has been reported that weeds are much less a problem in multicropping systems, especially in intercropping, because the space normally available to weeds is filled with other crops. The aggressive nature of weeds is well known, but recent work has begun to show that weeds can fill an important ecological role in cropping systems, by capturing unused nutrients, protecting the soil, altering soil fauna and flora, serving as trap plants for pests and disease, and changing the microhabitat to allow for high populations of pest predators and parasites. In rural tropical Mexico, farmers understand and use a "non-weed" concept, where each is classified according to positive or negative effects. We need to understand in more detail the biological functions of each component of the agroecosystem to establish the structure that will allow adequate weed, pest, and disease control. If part of this control can be achieved by merely manipulating the crop mixture in time and space, great strides toward more efficient agricultural management can be made.

5. Mutualisms and Crop Coexistence: In natural ecosystems, a great number of interactions between different species are mutually beneficial for those organisms involved, leading us to believe that there is a strong selective pressure operating to select combinations that coexist rather than compete. On the long term, such a coexistence permits a more efficient use of resources, with the component organisms aiding one another rather than interacting negatively. This frees more energy for growth and reproduction.

To a certain extent, nurse crops or companion plants function in this way. Legumes, because of their symbiosis with nitrogen-fixing bacteria, can coexist with corn without competing for nitrogen. In fact, part of the legume's nitrogen may be available for the corn, reducing overall need for fertilizers. Studies with coffee and cocoa shade trees have demonstrated the same relationship; the trees provide shade, nitrogen, and an organic mulch over the soil.

As mentioned, the presence of one crop may have beneficial effects on others through alteration in the microclimate, pest and disease protection, etc. Thus, apart from looking for crops that complement one another by avoiding overlap in requirements, we need also to look for crops that are interdependent and that mutually benefit from the association. This will be a very stimulating challenge for crop selection programs.

Use of Space and Time: One of the most important aspects of the management of multicropping systems is the facility they offer for the intensification of production through manipulation of space and time. By achieving the most ideal combination of the two, we will achieve the greatest productivity. On the one hand, we attempt to occupy the available resource space as efficiently as possible, combining species that complement each other, yet attempting to avoid overlaps that lead to negative interactions.

Resource use in space is then combined with its use in time, trying to achieve constant use of the resources available. For this reason, multicropping systems are intensified by sequential, relay, and mixed planting that establish constant resource use within the environmental limits imposed by the ecological conditions of each region. In this sense, we can even visualize the possibility of including cold resistant trees in association with annual crops or pasture, so that during the winter the trees continue to occupy the area. Thus, any yield reduction during the normal frost-free growing season is compensated for by the long-term tree production.

Additionally, multicropping systems permit greater stability in production, despite variability in climate or physical factors in the planting area. Whatever the conditions in one location and for one growing season, at least one member of the multicropping system will succeed. Since most of the better drained and structured soils are already in production, the more marginal lands will require special technology to make them produce. We cannot consider for the moment massive programs of soil and water manipulation needed to install mechanized high-yielding monoculture. To do so is economically, if not ecologically, prohibitive. The basic framework is available in multicropping, Innovative combinations need to be searched for and tested.

Agroforestry: A Multicropping System

Agroforestry is a technology of land management that combines trees with agricultural crops, with animals, or any combination of the two. Combinations can be simultaneous, or staggered in either time and space. The major objective of agroforestry is to optimize production for each unit of surface area, keeping in mind the need to maintain long-term yield. Small-scale, traditional agriculture has always included trees as integrated elements of farm management, but only recently has interest been revitalized in the application of agroforestry practices into modern agriculture.

The renewal of interest in agroforestry is based on many of the same reasons for multicropping systems in general: the ever increasing demand for production, yet the rising cost of obtaining it. The explosive demand for firewood and lumber has placed incredible pressures on the world's forests, especially in tropical and subtropical regions. Deforestation continues at an accelerated rate. But programs of reforestation or multiple-use forest management do not satisfy basic needs for food, clothing, and other necessities that come from crop and range lands. It would seem logical that these pressures for both forest and agricultural products would stimulate their combination in agroforestry systems.

Agroforestry practices can be broadly classified into three types combined agrosilvicultural (crop plus trees) systems, combined forestry and grazing simultaneous combinations of forestry with crops and grazing. Examples of each of these classifications are presented. The focus varies from soil improvement, erosion control, wind breaks, and shade to lumber, firewood, and reforestation. The combinations are essentially unlimited, depending on the needs of each region. At first glance it might appear that agroforestry systems are most applicable on marginal lands, on steep slopes, poor soils, or areas with widely fluctuating rainfall regimes. But agroforestry should also be considered for widespread application, even on prime agricultural or grazing land, because production needs to be increased—both by opening up new areas and by looking for innovative ways to increase productivity of lands already in use.

The principle limitations to widespread use of agroforestry practices are economic and technological. Ecologically, the advantages are well known, but technically we still do not have the information necessary to begin immediate implementation. With the present focus in agriculture aimed at maximizing single crop yields, there is a lack of acceptance of the idea that yields need to be thought of more on a long term, diversified basis. Agricultural research has not yet accepted the challenge that an integrated focus to forest and farm management requires.

Socioeconomic Implications of Multicropping Systems

In all of the aspects of multicropping systems that this review has considered—yield, resource use, pest and disease control, weeds, use of space and time, types of planting systems—much of the evidence indicates that generally there are more advantages than disadvantages of a biological, physical, or agronomic nature. But we need to consider the social and economic implications of the possible more widespread use of multicropping systems in present day agriculture.

The types of advantages derived from multicropping are many and varied. With a greater diversity of crops, a farmer is less affected by market fluctuations and is able to shift from one crop to another depending on price and demand. At the same time, the harvest is spread out over a longer period of time. Less dependence on outside energy sources has obvious advantages, especially in areas where capital is limited. Labor, instead of being concentrated in certain periods of the year, can be more evenly distributed, an important consideration in relation to the migrant farm worker problem. In times of higher unemployment, multicropping systems can offer more and steadier work.

Classification and Examples of Agroforestry Technologies

- Combined Agrosilvicultural Systems (Trees with Crops):

    ○ Agrosilviculture—establishment of trees, intercropped with agricultural crops during initial stages of tree growth, until tree canopies close and force the elimination of the crops. Production available in early stages of tree development, and cultivation activities simultaneously benefit both crops and trees.

    ○ Forest trees of commercial value in crop systems. Maintain trees in crop areas, either planted or natural, at low densities that do not interfere, yet provide value in the future.

    ○ Fruit trees in crop systems. A system that allows fruit production and grain or vegetable production simultaneously.

    ○ Trees that serve as shade for certain crops or improve the soil through nitrogen fixation, organic matter incorporation, mulch, and microclimate modification.

   ◦ Trees used as hedgerows, fence lines, or windbreaks around cropping areas, where management is intimately linked with the needs of the crops.

   ◦ Trees around rivers, lakes, or artificial reservoirs or tanks, integrated with fish or waterfowl management, providing shade, food, and roosting.

- Combined forestry and Grazing Systems (Trees with Grasses):

   ◦ Grazing or forage production takes place within forestry plantations, aiding in avoiding weed or brush build up, lowering fire risk.

   ◦ Grazing or forage production in young natural forests, with same advantages as above.

   ◦ Forest trees of commercial value in pastures, either planted or natural, at densities that do not interfere with the pasture species.

   ◦ Timber trees in pasture, either planted or natural, with the capacity to fix nitrogen and improve soil, thus lowering the need to fertilize and provide commercial value.

   ◦ Trees in pastures that provide shade for the animals and aid in improving the soil through nitrogen fixation and nutrient extraction from deeper soil levels.

   ◦ Trees, either in or around pastures, or in forests, that produce foliage of forage value for animal consumption. Can allow the reduction of feed supplement for animals.

   ◦ Fruit trees in pastures, allowing for commercial production of both fruits and animals.

   ◦ Trees around pastures as hedgerows, fence lines, or windbreaks.

- Simultaneous Combinations of Forestry with Crops and Grazing:

   ◦ Forest plantations planted with crops and grasses, permitting the management of grazing animals, either free to wander or enclosed in specific areas. Especially adapted to smaller animals, such as ducks or pigs. Requires close control of activities and use of specific crops.

   ◦ Trees associated with crops and grazing, either planted or natural, in densities that will not adversely influence the crops. Trees scattered in and around cropping areas can be periodically pruned and used as forage for animals, with the timber harvestable at some later date.

   ◦ Hedgerows or living fence lines around rural communities, serving as shade, windbreak, property divisions, forage, fruits, timber, and firewood. In this sense, the system is truly multiple use.

Most of the economic disadvantages are derived from our lack of experience and knowledge with multicropping systems. Reported lower yields, complexity of management activities, higher labor demands, and the difficulty in mechanizing such systems are all important factors that discourage modern farmers from participating in multicropping practices.

An important aspect of this resistance comes from the emphasis on large profits that governs so much of modern agriculture today. Maximum profits in the short term, rather than concern with maintaining constant income in the long term, governs the decision making process on most American farms today. But with the incredible rise in farm costs, a new focus is necessary. All of these increases cannot be passed onto the consumer. Many of the advantages of multicropping systems definitely need to be stressed more for use on farms today. Smaller farms, with a greater diversity of products and activities, can function quite profitably because they are less dependent on high-cost energy inputs. Lower costs mean food can be produced at a lower price, the benefits being transferred to the general population.

Smaller farms would require more farmers. To a certain extent multicropping systems mean a return to the land, with the incentives necessary to keep the farmers there. The great diversity of activities in multicropping systems would promote an increase in interdisciplinary activities in their investigation, installation, management, and use in agriculture. This stimulation of interchange and collaboration can, in the long term, lead to greater social cohesion. Rural regions might once again take on the social importance they enjoyed in the past. The problems of lack of trained personnel, and social, political, and economic restrictions on multicropping systems, all can be overcome by thorough and conscientious programs of research aimed at determining the proper methods, varieties, and practices necessary.

The belief that multicropping is only suitable for marginal or underdeveloped regions ignores the fact that just a relatively short time ago, such systems were the most common type of agriculture. Only recently have they been replaced by monoculture systems dependent on the use of massive quantities of inexpensive high energy inputs. For the moment, this time has passed and we need to learn from the past to reshape agriculture for the future. This will be a great challenge for agricultural research.

Types, Advantages and Disadvantages of Multicropping

Types of Multicropping

There are many types of multicropping system:

- Intercropping,
- Mixed cropping,
- Relay cropping,

- Sequential cropping,
- Ratoon cropping, etc.

Advantages of Multicropping

Multicropping systems are not only beneficial for soil, but it is also beneficial for farmer as well as for the country.

- It is a better source of soil utilization,
- It improves yield,
- Increase yield per unit of land,
- Increase the crop production,
- Export increase,
- Foreign exchange increase,
- Costs of input decrease as compared to individual crop growing cost,
- Reduce pest and disease attack,
- Different type of products can be produce at a time,
- It helps to produce a balance diet for a family,
- It helps to maintain the soil fertility,
- It helps to control weeds.

Disadvantages of Multicropping

Multicropping systems also have some disadvantages like:

- The survival of pests become easy,
- Pests can easily shift from one crop to another crop,
- Problem of weed management,
- Implementation of new technology is difficult etc.

Intercropping

Sustainable agriculture seeks, at least in principle, to use nature as the model for designing agricultural systems. Since nature consistently integrates her plants and animals into a diverse landscape, a major tenet of sustainable agriculture is to create and maintain diversity. Nature is also efficient. There are no waste products in nature. Outputs from one organism become inputs for another. One organism dies and becomes food for other organisms. Since we are modeling nature, let us first look at some of the principles by which nature functions. By understanding these principles we can use them to reduce costs and increase profitability, while at the same time sustaining our land resource base.

Diversity is Nature's Design

When early humans replaced hunting and gathering of food with domestication of crops and animals, the landscape changed accordingly. By producing a limited selection of crop plants and animals, humankind has greatly reduced the level of biological diversity over much of the earth.

Annual crop monocultures represent a classic example. In response to this biological simplification, nature has struggled to restore diversity to these landscapes—that is her tendency. Our "war" with nature over the tendency to diversity is what we call "weed control" and "pest management." Of course we could hardly produce any crops if we simply allowed our fields to return to natural vegetation, but we can realize some of the benefits of diversity by planting mixtures of different crops.

Cooperation is more Apparent than Competition

There is far more cooperation in nature than competition. Cooperation is typified by mutually beneficial relationships that occur between species within communities. In The Redesigned Forest, ecologist Chris Maser offers a glimpse of the cooperation inherent in a northern temperate forest when he describes a relationship that exists among squirrels, fungi, and trees. The squirrels feed on the fungus, then assist in its reproduction by dropping fecal pellets containing viable fungal spores onto the forest floor. There new fungal colonies establish. Tree feeder roots search out the fungi and form a symbiotic association that enables the tree roots to increase their nutrient uptake. The fungi, in turn, derive food from the tree roots. Each benefits from the other's presence or actions.

If we view competition as the driving force in nature, we fail to see the complex relationships and feel compelled to take actions that may have unforeseen impacts. The rancher who views coyotes as competitors (for calves and lambs) and kills them out may later find the predator helped keep rodent populations in check. With the predator gone, rodent numbers explode and cause more problems than ever before. The same is true with many insect pests of crops. When the only food for insects is crops, that is what they will eat. With no predator or parasite habitat present in a pure stand of crop, the pest insect could not possibly have it better. If we can shift our view of nature from a theme of competition to one of collaboration, we can act in ways that yield fewer negative consequences.

Stability Tends to Increase with Increasing Diversity

If left undisturbed and unplanted, an abandoned crop field will first be colonized by just a few species of plants, insects, bacteria, and fungi. After several years, a complex community made up of many wild species develops. Once a wild plant and animal community has reached a high level of diversity, it remains stable for many years.

When wild communities are in the early stages of development, or when they have lost diversity due to natural catastrophe or human actions, they are prone to major fluctuations, both in types of species present and in their numbers. Disease outbreaks in plants and animals occur more frequently—as do outbreaks of weed, insect, bird, or rodent pests. One good example is the grasshopper plagues that follow regional weather shifts. Another is the shift in weed species dominance following a soil disturbance.

The more complex and diverse communities become, the fewer the fluctuations in numbers of a given species, and the more stable communities tend to be. As the number of species increases, so does the web of interdependencies. In both higher and lower rainfall years, there are fewer increases in any one species and fewer fluctuations in the community as a whole.

Pursuing Diversity on the Farm

So, then, how can we begin to model our agricultural pursuits after some of these natural principles? Can we look for patterns in nature and imitate them? Some pioneering farmers have been able to utilize nature's principle of diversity to their advantage. Results of their efforts include lower cost of production and higher profits. Among the practices that promote diversity and stability are:

- Enterprise Diversification: Risk reduction through stability of income and yield are two of the reasons people diversify their crop and livestock systems. Increasing diversity on-farm also reduces costs of pest control and fertilizer, because these costs can be spread out over several crop or animal enterprises.

- Crop Rotation or Rotation: Moving from simple monoculture to a higher level of diversity begins with viable crop rotations, which break weed and pest life cycles and provide complementary fertilization to crops in sequence with each other.

- Farmscaping: Diversity can be increased by providing more habitat for beneficial organisms, habitats such as borders, windbreaks, and special plantings for natural enemies of pests. Request the ATTRA publication Farmscaping to Enhance Biological Control for more information on special plantings for beneficial insects.

- Intercropping: Intercropping is the growing of two or more crops in proximity to promote interaction between them. Much of this publication focuses on the principles and strategies of intercropping field crops. A related ATTRA publication, Companion Planting, provides more information on intercropping of vegetable crops.

- Integration: On-farm diversity can be carried to an even higher level by integrating animals with intercropping. With each increase in the level of diversity comes an increase in stability. This publication focuses on intercropping and provides a section on integrating livestock with crops.

Intercropping Concepts

Most grain-crop mixtures with similar ripening times cannot be machine-harvested to produce a marketable commodity since few buyers purchase mixed grains. Because of limited harvest options with that type of intercropping, farmers are left with the options of hand harvesting, grazing crops in the field with animals, or harvesting the mixture for on-farm animal feed. However, some intercropping schemes allow for staggered harvest dates that keep crop species separated. One example would be harvesting wheat that has been interplanted with soybeans, which are harvested later in the season. Another example is planting harvestable strips, also known as strip cropping.

When two or more crops are growing together, each must have adequate space to maximize cooperation and minimize competition between them. To accomplish this, four things need to be considered:

- Spatial arrangement,

- Plant density,

- Maturity dates of the crops being grown, and

- Plant architecture.

Strip cropping.

Spatial Arrangement

There are at least four basic spatial arrangements used in intercropping. Most practical systems are variations of these.

- Row intercropping: Growing two or more crops at the same time with at least one crop planted in rows.

- Strip intercropping: Growing two or more crops together in strips wide enough to permit separate crop production using machines but close enough for the crops to interact.

- Mixed intercropping: Growing two or more crops together in no distinct row arrangement.

- Relay intercropping: Planting a second crop into a standing crop at a time when the standing crop is at its reproductive stage but before harvesting.

Plant Density

To optimize plant density, the seeding rate of each crop in the mixture is adjusted below its full rate. If full rates of each crop were planted, neither would yield well because of intense overcrowding. By reducing the seeding rates of each, the crops have a chance to yield well within the mixture. The challenge comes in knowing how much to reduce the seeding rates. For example, if you are planning to grow corn and cowpeas and you want mostly peas and only a little corn, it would be easy to achieve this. The corn-seeding rate would be drastically cut (by 80% or more) and the pea rate would be near normal. The field should produce near top yields of peas even from the lower planting rate and offer the advantage of corn plants for the pea vines to run on. If you wanted equal yields from both peas and corn, then the seeding rates would be adjusted to produce those equal yields.

Maturity Dates

Planting intercrops that feature staggered maturity dates or development periods takes advantage of variations in peak resource demands for nutrients, water, and sunlight. Having one crop mature before its companion crop lessens the competition between the two crops. An aggressive climbing bean may pull down corn or sorghum growing with it and lower the grain yield. Timing the planting of the aggressive bean may fix the problem if the corn can be harvested before the bean begins to climb. A common practice in the old southern U.S. cotton culture was to plant velvet beans or cowpeas into standing corn at last corn cultivation. The corn was planted on wide 40-inch rows at a low plant population, allowing enough sunlight to reach the peas or beans. The corn was close enough to maturity that the young legumes did not compete. When the corn was mature, the beans or peas had corn stalks to climb on. The end result was corn and beans that would be hand harvested together in the fall. Following corn and pea harvest, cattle and hogs would be turned into the field to consume the crop fodder.

Selecting crops or varieties with different maturity dates can also assist staggered harvesting and separation of grain commodities. In the traditional sorghum/pigeonpea intercrop, common in India, the sorghum dominates the early stages of growth and matures in about four months. Following harvest of the sorghum, the pigeonpea flowers and ripens. The slow-growing pigeonpea has virtually no effect on the sorghum yield.

Plant Architecture

Plant architecture is a commonly used strategy to allow one member of the mix to capture sunlight that would not otherwise be available to the others. Widely spaced corn plants growing above an understory of beans and pumpkins is a classic example.

Intercrop Productivity

One of the most important reasons to grow two or more crops together is the increase in productivity per unit of land. Researchers have designed a method for assessing intercrop performance as compared to pure stand yields. In research trials, they grow mixtures and pure stands in separate plots. Yields from the pure stands, and from each separate crop from within the mixture, are measured. From these yields, an assessment of the land requirements per unit of yield can be determined. This information tells them the yield advantage the intercrop has over the pure stand, if any. They then know how much additional yield is required in the pure stand to equal the amount of yield achieved in the intercrop. The calculated figure is called the Land Equivalency Ratio (LER). To calculate an LER, the intercrop yields are divided by the pure stand yields for each component crop in the intercrop. Then, these two figures are added together. Here's the equation for a corn/pea intercrop where the yields from pure corn, pure peas, and the yields from both corn and peas growing together in an intercrop are measured.

When an LER measures 1.0, it tells us that the amount of land required for peas and corn grown together is the same as that for peas and corn grown in pure stand (i.e., there was no advantage to intercropping over pure stands). LERs above 1.0 show an advantage to intercropping, while numbers below 1.0 show a disadvantage to intercropping. For example, an LER of 1.25 tells us that the yield produced in the total intercrop would have required 25% more land if planted in pure stands. If the LER was 0.75, we know the intercrop yield was only 75% of that of the same amount of land that grew pure stands.

In a South Carolina study, researchers planted intercrops of southern peas and sweet corn at three different corn plant densities. The plantings were on raised beds with flat and wide crowns on six-foot centers. In the center of each bed was a corn row, with two rows of peas planted 18 inches to either side of the corn row. The low corn-seeding rate was 6,700 plants per acre, medium corn was 9,500 per acre, and high was 11,900 plants per acre. Peas were established at a rate of 31,800 plants per acre in all intercrop plots. In the pure pea stand, each bed had two rows of peas spaced 24 inches apart. Yields of the intercrops and pure stands are shown in table.

In this trial there was a yield advantage from intercropping over growing the two crops in pure stands. Pea yields suffered from the increased competition in the higher densities of corn. Some practical on-farm guidelines can be drawn to guide seeding-rate choices for a two-crop intercrop. To test seeding rates, experiment with three small plantings of two crops at the following percentages of their full seeding rates: 1/3 + 2/3, 1/2 + 1/2, and 2/3 + 1/3. From there, make adjustments for future plantings based on the results and your expectations.

Table: Yields of sweet corn and southern peas from intercrops.

Seed Rates	Corn (pounds/ acre)	Peas (pounds /acre)	LER
Full corn	5600	***	***
Full peas	***	1200	***
Low corn	4200	800	1.41
Medium corn	4600	800	1.48
High corn	5000	500	1.30

Managing Intercrops

Many combinations of crops have been grown or experimented with as mixed or relay intercrops. Some of these include sunflowers grown with black lentils, wheat with flax, and canola with flax. Other combinations include cucumbers, beans, celery, and chives in China; upland rice, corn, and cassava in Indonesia; and in various parts of the tropics corn and cassava, corn and peanuts, sorghum and millet, and sorghum and pigeonpeas.

Sweetcorn and southern pea planting pattern.

Frequently these cropping combinations involve a short and a tall crop both planted at the same time. In many cases the tall crop is harvested first. For example, corn grown with a shorter plant would be harvested first, then peanut or sweet potato would be harvested later. Another pattern would be planting two tall crops with different growth rates. In relay intercrops, different planting dates are used so that one crop might mature sooner. Corn or sorghum, requiring three months to mature, can be grown with pigeon pea, requiring 10 months to maturation.

John Bowen and Bernard Kratky, researchers and instructors at the University of Hawaii, tell us that there are five distinct aspects to successful multicropping. These are:

- Detailed planning,

- Timely planting of each crop,

- Adequate fertilization at the optimal times,

- Effective weed and pest control,

- Efficient harvesting.

Before any fieldwork is begun, adequate planning should be done. Planning covers selection of crop species and appropriate cultivars, water availability, plant populations and spacing, labor requirements throughout the season, tillage requirements, and predicted profitability of the intercrop. These and other parameters need to be evaluated before spending money on inputs.

With any crop, seed germination and seedling establishment are the most critical phases of the entire season. A good seedbed is needed to get a good stand. Delayed planting may reduce yield, since crop development may not coincide with the optimal growth periods.

Planning fertilization for intercrops can be challenging, as the full needs of both crops must be met. Generally, there is little information available on how to go about this. One possibility would be to ask for soil test results for each crop separately, then formulate a recommendation that will cover the needs of both crops to be grown. Such recommendations are generally 10% to 30% higher than rates for individual crops.

As with any crop, also accounting for residual or carryover fertility from past crops saves money. Carryover fertility from intercrops may well be lower than that of pure stands because of the two crops having different root types and feeding habits.

Weed and pest controls in intercrops will likely be different from those in pure stands. Some disease incidence, such as soybean or mung bean rusts, may increase when aggravated with high corn populations and over fertilization. Any disease or pest that prospers in shady conditions could increase under a taller crop such as corn or sunflowers. In many cases, insect pest populations are lower when two or more crops are grown together. More on pest management will be found later in this publication.

Harvesting of mixed intercrops has been a major limitation to their adoption in mechanized farming. As mentioned earlier, if the crops cannot be harvested by animals, or all together as feed, you're left with hand harvesting. Some crops such as flax and wheat have been harvested together and mechanically separated. Any other mechanized harvest efforts must get one crop without damaging the other. One example would be harvesting wheat over the top of a young stand of soybeans growing beneath the grain heads. All intercropping strategies— especially mixed intercropping—require advanced planning and keen management. Success will likely be the reward for such efforts.

Examples of Intercrop Systems

Traditional Corn-bean-squash Mixed Intercrops

Farmers throughout Central America traditionally grow an intercrop of corn, beans, and squash. Grown together, these three crops optimize available resources. The corn towers high over the other two crops, and the beans climb up the corn stalks. The squash plants sprawl along the ground, capturing light that filters down through the canopy and shading the ground. The shading discourages weeds from growing.

This mixture was compared to the individual crops grown separately in a study near Tabasco, Mexico. In the study, corn yields were considerably higher in the mixture than in a pure stand planted at optimum densities. Bean and squash yields suffered considerable yield reductions when grown in mixture. In this example if corn were the most important crop, it was beneficial to grow it in a mixture with squash and beans. The beans and squash were just a bonus. The LER for the whole mixture was considerably higher than any of the pure stands.

Yields of corn, beans and squash grown alone or in a mixture		
Crop	Pure Stand (pounds/acre)	Intercrop (pounds/acre)
Corn	1096	1533
Beans	544	98
Squash	383	71

Corn and Soybean Mixed Intercrops

Canadian researchers have worked with several corn-soybean intercrop seeding rates to determine their economic advantages as silage. Pure stands of corn and soybeans were grown for comparison at 24,000 corn seed per acre and 200,000 soybean seed per acre. Results showed that intercrops were more cost effective than pure stands over both years the study was conducted. The study featured five experimental intercrop seeding rates with two planting arrangements (alternate and within the row). The researchers concluded that a planting rate of 16,000 corn seed per acre (67% of the full corn rate) with 135,000 soybean seed per acre (67% of the full bean rate) planted within the same rows along with 53 lbs. of N/acre gave the highest economic returns. (The planter was set to drop 151,000 seeds per acre.) This mixture gave an LER of 1.14 over pure stand yields. The crude protein level of the intercrop silage was considerably higher than that of pure corn silage. A slightly higher yield was achieved from full stands of both corn and beans in alternate rows (LER=1.23), but the cost of production was higher, thus offsetting the improved yields.

Corn and Sorghum Mixed Intercrops

Frank Cawrse, Jr., of Lebanon, Oregon, intercrops forage sorghum into his silage corn. He first plants the corn at 28,000 seed per acre, then goes back over the field with a drill with enough drop tubes closed off to plant 8 pounds of sorghum on 32-inch rows in between the corn. He also plants two different maturities of corn, a 95- day and a 75-day, to even out the silage moisture content. He harvests a mix of corn in hard dent and soft dent, and sorghum in the milk stage.

Strip Cropping Corn/Soybeans/Small Grains

South Dakota farmer Tod Intermill plants alternating strips of corn, soybeans, and spring wheat on his farm. The strips are six rows wide in a ridge-till system. All the crop plantings are adapted to existing equipment widths. Regular herbicide treatments can be applied using a ground sprayer of strip width. Even the wheat is drilled on ridges, using a drill with individual depth gauges on each opener. Intermill orients his rows east and west to minimize the shading effects of taller crops like corn. The crops are planted in a wheat–corn– soybean pattern, with soybeans on the north side of the corn.

This arrangement reduces the effect of corn shading often associated with a straight corn-soybean pattern, since the wheat is mature before the corn has a chance to shade it. Corn gains the greatest benefit from the additional sunlight interception on the outside rows of the corn strip.

Corn, soybeans, and wheat stripcropped.

Iowa farmer Tom Frantzen strip-crops oats, corn, and soybeans on ridge-till rows. He views his strips as a crop rotation in one field. His rows are oriented generally east and west on the contour. His 1989 strip-crop corn yields were 166 bushels per acre, compared to 130 for his farm average. Stripped soybean yields were two bushels lower than farm average. His oat yields were 109 bushels stripped and 100-bushel farm average. Tom was not surprised at the increase in corn yields. The outer strip rows captured more sunlight. His average corn border row yielded 198 bushels per acre next to the soybeans and 177 bushels next to oats. The soybean yields were 37 bushels, even with the increased shading on the border rows. This loss was made up in the middle rows with yields of 44 bushels per acre. Oats showed a 107-bushel yield on the soybean side, a 103-bushel yield on the corn side, and 99 bushels in the middle. Tom says the strip intercropping is no more labor intensive than monocrop fields. His profits were $76 per acre for the stripped fields and $55 for the same crops grown in monoculture.

Rick Cruse, an Iowa State University agronomist, has observed several characteristics that narrow strips (12 to 30 feet wide) offer. The strips accommodate the pest management and soil building advantages of rotations and the yield boost of border rows. With proper management the border effect can pay off; managed improperly, it can cost yield. With oat and corn strips, the early-maturing oats are nearly mature before corn can pose much of a shade and competition problem. The corn can also provide wind protection for the oats. When the oats are harvested, more sunlight is available to the corn. In times of low moisture, oats may rob the corn border rows of water. In his field trials, Cruse found a 5% increase in oat yields on their borders, while corn realized a 12 to 15% increase.

Soybean yields dropped by 10% on their border rows, but the yields in the soybean middle rows were higher than they would be in a solid field, possibly representing a windbreak effect.

Some have experimented with a shorter corn variety in the border row to minimize shading. One farmer tried planting six rows of corn and doubling his soybean strips to 12 rows to eliminate the impact of corn shading on the beans. This same farmer found that corn strips wider than eight rows did not provide adequate results. Using a 12-row planter, it's easy to establish the 6-row strips by filling the middle six hoppers with corn and the outer three hoppers with beans. Some farmers plant higher corn populations and add higher nitrogen rates in the border rows to take advantage of the extra sunlight exposure. Most farmers agree that strip cropping corn, soybeans, and oats works best with ridge-till or no-till. When the field is tilled, it's difficult to gauge where the rows should go in order to get the strips even.

Escalating Diversity and Stability to a Higher Level

Ecologists tell us that stable natural systems are typically diverse, containing many different types of plants, arthropods, mammals, birds, and microorganisms. In stable systems, serious pest outbreaks are rare, because natural controls exist to automatically bring populations back into balance. Planting crop mixtures, which increase farmscape biodiversity, can make crop ecosystems more stable, and thereby reduce pest problems.

There is overwhelming evidence that plant mixtures support lower numbers of pests than do pure stands, and there are two schools of thought on why this occurs. One suggests that higher natural enemy populations persist in diverse mixtures due to more continuous food sources (nectar, pollen, and prey) and favorable habitat.

The other thought is that pest insects that feed on only one type of plant have greater opportunity to feed, move around in, and breed in pure crop stands because their resources are more concentrated than they would be in a crop mixture. Regardless of which reason you accept, the crops growing together in the mixture complement one another, resulting in lower pest levels.

Intercropping also aids pest control efforts by reducing the ability of the pest insects to recognize their host plants. For example, thrips and white flies are attracted to green plants with a brown (soil) background, ignoring areas where vegetation cover is complete—including mulched soil. Some intercrops have a spatial arrangement that produces the complete vegetation cover that would be recognized as unfavorable to thrips and whiteflies. Other insects recognize their host plants by smell. Onions planted with carrots mask the smell of carrots from carrot flies. For more information on companion planting for insect management, request the ATTRA publications Farmscaping to Enhance Biological Control and Companion Planting.

Innovative farmers are paving the way with intercrops and realizing pest management benefits as a result. Georgia cotton farmers Wayne Parramore and sons reduced their insecticide and fertilizer use by growing a lupine cover crop ahead of their spring-planted cotton. They started experimenting with lupines on 100 acres in 1993, and by 1995 were growing 1,100 acres of lupines. Ground preparation for cotton planting is begun about 10 days prior to planting by tilling 14-inch wide strips into the lupines. Herbicides are applied to the strips at that time, and row middles remain untouched. The remaining lupines provide a beneficial insect habitat and also serve as a smother crop to curtail weeds and grasses. The lupines in the row middles can be tilled in with the cultivator later in the season to release more legume nitrogen.

Lupine

In the Parramores' system, all the nitrogen needs of the cotton crop are met with cover crops except for 10 units per acre of starter nitrogen and another 15 units applied while spraying herbicides. Petiole samples taken every week to monitor plant nitrogen show that cotton grown with lupines maintains a normal range of tissue nitrogen throughout the growing season. The nitrogen level in cotton grown solely with fertilizer is very high initially, then subsequently falls back to a lower level. In one comparative year, the cotton grown following lupine produced 96 more pounds of lint, with only 25 units of commercial nitrogen, compared to a field with 125 units of nitrogen and no lupines. Additionally, the lupine field required less spraying for insects—only twice compared to five sprays for the commercial nitrogen field. This reduction saved 60% on insecticides, amounting to $35 per acre. The reduction in need for pesticides is attributed to the large population of beneficial insects generated and sustained in this system. The lupines provide food for aphids and thrips, which attract ladybugs, big-eyed bugs, and fire ants as predators. When the cotton gets big enough to shade out the lupines, the beneficial insects move to the cotton rather than migrating from the field. The Parramores estimate that improved yields, combined with cost reductions, are netting them an additional $184 per acre with the strip tillage lupine system when compared to the conventional management system.

Alfalfa is one of the best crops for attracting and retaining beneficial insects. This characteristic can be enhanced further. Strip-cutting alfalfa (i.e., cutting only half of the crop at any one time, in alternating strips) maintains two growth stages in the crop; consequently, some beneficial habitat is available at all times. In some cases alfalfa is mixed with another legume and a grass. Auburn University researcher Mike Gayler is just starting research projects using alfalfa as an attractant crop for beneficials. He speculates that it will work in the Southeast with proper management. Other main-sea-son strip crops that research suggests will benefit cotton crop pest management include cowpeas, sorghum, corn, and crotalaria.

Dr. Sharad Phatak of the University of Georgia has been working with cotton growers in Georgia testing a strip-cropping method using annual winter cover crops. Planting cotton into strip-killed crimson clover improves soil health, cuts tillage costs, and allows him to grow cotton with no insecticides and only 30 pounds of nitrogen fertilizer. Working with Phatak, farmer Benny Johnson reportedly saved at least $120/acre on his 16-acre test plot with the clover system. There were no insect problems in the test plot, while beet armyworms and whiteflies were infesting nearby cotton and requiring 8 to 12 sprayings to control. Cotton intercropped with crimson clover yielded more than three bales of lint per acre compared to 1.2 bales of lint per acre in the rest of the field. Boll counts were 30 per plant with crimson clover and 11 without it. Phatak identified up to 15 different kinds of beneficial insects in these strip-planted plots.

Phatak finds that planting crimson clover seed at 15 pounds per acre in the fall produces around 60 pounds of nitrogen per acre by spring. By late spring, beneficial insects are active in the clover. At that time, 6- to 12-inch planting strips of clover are killed with Roundup herbicide. Fifteen to 20 days later the strips are lightly tilled and cotton is planted. The clover in the row middles is left growing to maintain beneficial insect habitat. When the clover is past the bloom stage and less desirable for beneficials, they move readily onto the cotton. Even early-season thrips, which can be a problem following cover crops, are limited or prevented by beneficial insects in this system. The timing coincides with a period when cotton is most vulnerable to insect pests. Following cotton defoliation, the beneficials hibernate in adjacent non-crop areas.

Phatak points out that switching to a whole farm focus while reducing off-farm inputs is not simple. It requires planning, management, and several years to implement on a large scale. It is just as important to increase and maintain organic matter, which stimulates beneficial soil microorganisms. Eventually a "living soil" will keep harmful nematodes and soilborne fungi under control. For more information on management of soil-borne diseases, request the ATTRA publication Sustainable Management of Soilborne Plant Diseases.

Texas dryland farmer Ron Gobel intercrops 8-row strips of sesame and cotton for insect control benefits. The sesame harbors many beneficial insects, including high populations of lacewings, assassin bugs, and lady beetles. Ron's 1995 crop was planted late due to prolonged spring rains. He did not use a Bt cotton variety. Early frost terminated the crop two weeks earlier than normal, yet he still produced 0.8 bales per acre under dryland conditions. His sesame produced 800 pounds per acre. The 1996 cotton rows were planted where the sesame rows were the previous year, and sesame planted where cotton was before.

Since Ron sells his cotton for a premium price in the organic market, he cannot spray any synthetic insecticides. Consequently, he must rely on beneficial insects attracted to his fields by cultural practices and a handful of natural insecticides.

Following the fall harvest, Ron plants annual rye at a low rate of 20 to 40 pounds per acre. The rye is tilled in prior to crop planting in the spring. Ron believes the rye helps with soil moisture retention and weed control. During the 1997 crop year his fields suffered only minimal boll weevil damage. Ron noticed lots of adult bollworm moths but no worms. The eggs were eaten or parasitized by the beneficials.

Ron's fields were scouted as part of a boll weevil eradication program. The scouts were amazed at the lack of worms and the high numbers of beneficial insects. The cotton crop was sprayed one time with diatomaceous earth impregnated with natural pyrethrum, which was acceptable under the organic standards. The insect scouts noticed a 70% reduction in adult boll weevil population three days after the spray. They were so surprised that they placed cages of 20 live weevils in the field to see whether the spray was working. The next day, 45% of those weevils were dead. The entomologists speculated that the weevils were getting enough of the diatomaceous earth on their leg joints to cut their exoskeletons, allowing the pyrethrum to kill them.

In a scientific study, Mississippi researchers interplanted 24 rows of cotton with 4 rows of sesame to study the intercrop's effects on tobacco budworms and bollworms (Heliothis spp.). Throughout the growing season, larvae numbers were much higher in the sesame than on the cotton until late August, indicating the worm's preference for sesame. Following a large summer rain at a time when the sesame was reaching maturity, the Heliothis adults became more attracted to the cotton. The researchers noted that sesame's attractiveness to Heliothis and sesame's ability to harbor high numbers of beneficial insects made it useful in a cotton pest management program.

Escalating Diversity and Stability to an Even Higher Level

The diversity created by intercropping can be enhanced even further by integrating livestock (single or mixed species) into the cropping plan as harvesters. Allowing animals to harvest feed crops in the field puts gain on animals at the cost of crop production—considerably less than the purchase price of the grain. If you think about it, feed grains cost a lot less when they're not run through a $150,000 combine or hauled 1000 miles across the country.

Grazing animals and other livestock can be managed on croplands to reduce costs, increase income, and increase diversity. There are ways of incorporating animals into cropping without the farmer getting into animal husbandry or ownership directly. Collaboration with neighbors who own animals will benefit both croppers and livestock owners. Grazing or hogging-off of corn residue is one example where a cost can be turned into a profit. The animals replace the $6 per acre stalk mowing cost and produce income in animal gains.

Shasta College provides a unique demonstration of integrating livestock with intercrops. Shasta is a two-year community college located in Redding, California, that offers associate degrees in several branches of agriculture. Stan Gorden heads the college's holistic resource laboratory, where students get hands-on experience with ranching and farming. Stan and his students have taken intercropping to a high level of efficiency. They run hogs, sheep, cattle, and chickens together over 42 small paddocks of various forages and crops growing on 100 acres of college-owned land. One paddock is a pumpkin patch, another a garlic and carrot patch. Some are planted in alfalfa or mixes of grasses and clover. Not all the pastures have water sources for the animals, so water is moved on a trailer tank when necessary. The animals are moved daily in a planned grazing system during rapid plant growth and much more slowly, up to five days on a paddock, during slow plant growth.

Some of the paddocks are planted with mixtures of either winter or summer forage or grain crops. An intercrop of cereal grain, fava beans, and Canadian field peas is planted for winter grain, each crop at 1/3 normal seeding rate. The grain mixture is combine-harvested to make energy and protein supplement feed as needed. After harvest, the animals are turned into the paddock to glean what's left. For summer feed, a mixture of milo planted on 18-inch rows is intercropped with a row of black-eyed peas planted six inches to either side of each sorghum row, using a drill with partitions in the seedbox. The milo provides a trellis for the pea vines to run on. The milo/black-eyed mixture requires no herbicide. Before peas and milo were grown together, the milo pure stand would be plagued with whiteflies and green bugs. Mixing the two crops together ended the pest problem. Cowpeas have extrafloral nectaries that attract lots of beneficial insects.

This could explain the absence of pest insects in the mixture. The milo/pea mixture is harvested by setting the combine to cut at the height of the milo heads. This yields a milo to bean ratio of 2:1—ideal for feed.

Cowpeas and milo growing together.

The college animal herd consists of 20 sows that farrow on pasture, 35 head of cattle, 50 sheep, and 30 laying hens that all range together. The hens are with the herd during the day and roost in a nearby egg mobile at night. Gorden selects breeds and genetics to fit this system, as opposed to selecting breeds for maximum production and adapting a system to match the animal. The animals benefit one another. The sheep learn to stay close to the middle of the herd to avoid predators, which are fended off by the hogs. The cattle learn that the hogs know how to break the pumpkins open, so they stick close and get some too. The hogs eat the cow and sheep droppings and benefit from the predigestion. The hens scavenge wasted seeds from the various crops. There are three different kinds of hens, each of which lays eggs of a different color. The eggs are marketed as rainbow eggs, with each dozen containing four white, four blue, and four brown eggs. The chickens also scratch apart cattle dung pats searching for insects, thus destroying cattle parasites.

Gorden says that developing and maintaining this high level of diversity has required creativity, selection criteria, constant monitoring, and re-examining traditional beliefs. By challenging long-held beliefs, Bill and his students discovered that hogs do not need standard farrowing crates and that sheep and cattle are compatible grazers. Animals and crops are selected and culled according to their ability to adapt to this complex system. Shasta College has one of the largest heritage hog herds in the country. The hogs have been fitted with humane nose rings to prevent rooting. Also, hog breeds are selected that don't root up the ground nor eat the baby lambs when they are born. The sows farrow on pasture with only a single bale of hay for bedding. Hogs are not vaccinated, nor are needle teeth removed or other detailing done. Sows generally wean 12 pigs with no supplemental feed. The only purchased input is some nitrogen and phosphorus fertilizer applied to the pastures. The pigs are only touched twice; once to castrate and once to wean. As with the hogs, the cattle and sheep are selected to prosper on grass. Predators are not controlled in any way. Any animal that gets killed by wandering off is naturally selected out of the herd.

The sheep/hog/cow mix provides much better utilization of forage than single species grazing. Since the animals do most of the harvesting, less fossil fuel and labor-hours are expended. There are no pens to wash and no manure to deal with. The herd is controlled using an electric fence charged up to 8,000 volts to hold the sheep.

Before the 100-acre crop/animal integration project began in 1987, the College's agriculture resource laboratory was costing $8,000 per year. That was the first year the resource laboratory started managing holistically. By 1996, the resource lab's income was up $12,000, and expenses were down $10,000—rendering a $14,000 profit over the 1987 figure. During that same time the soil organic matter has increased from 1.7% to 3.2%.

Intercropping for Disease Control

Under direction of an international team of scientists, farmers in China's Yunnan province made some simple changes in their rice production methods. They changed from planting their typical pure stand of a single rice variety to planting a mixture of two different rice varieties. Their primary reason for trying this new technique was to reduce the incidence of rice blast, the main disease of rice. The technique was so successful at reducing blast disease that the farmers were able to abandon chemical fungicides they had been using. The biodiversity effect is apparent here in that if one variety of a crop is susceptible to a disease, the denser the stand, the worse the disease can spread. If susceptible plants are separated by non-host plants that can act as a physical barrier to the disease, the susceptible variety will suffer less disease infection. Rice blast moves from plant to plant via airborne spores. These spores can be blocked by a row of a resistant variety. In this on-farm study, the rice was harvested by hand. Separating the varieties was easily done during harvest, since one variety towered above the other.

Adapting Intercropping to your Farm

Intercropping has been important in the U.S. and other countries and continues to be an important practice in developing nations. In traditional systems, intercropping evolved through many centuries of trial and error. To have persisted, intercropping had to have merit biologically, environmentally, economically, and sociologically. To gain acceptance, any agricultural practice must provide advantages over other available options in the eyes of the practitioner. Many of the impediments to adoption of new strategies or practices of diversification are sociological (Will I look foolish to my neighbors? Will I fail?) and financial (What are the risks? What is the profit potential?) rather than technological.

Farmers have generally regarded intercropping as a technique that reduces risks in crop production; if one member of an intercrop fails, the other survives and compensates in yield to some extent, allowing the farmer an acceptable harvest. Pest levels are often lowered in intercrops, as the diversity of plants hampers movement of certain pest insects and in some cases encourages beneficial insect populations.

Advantage and Disadvantage of Intercropping

Intercropping is quite common around the world and it is not just farmers but also hobbyists and gardeners who practice intercropping. There are advantages and disadvantages of intercropping. You should consider the rewards and challenges before you decide either way.

List of Advantages of Intercropping

Greater Income and Greater Yield

Intercropping offers greater financial returns for a farmer. Even if you are growing some produce for your own family or just as part of a hobby, you will have multiple types of produce, which is always a nice outcome. Intercropping will help farmers use the same land as available and yield more as well as diverse produce. This generates more income for the farmer without really taking up any major expenditure. The infrastructure available or the land used remains the same.

Insurance against Crop Damage

Intercropping can be the insurance that farmers need, especially when the region is vulnerable to weather extremes. Drought, torrential rain, hurricanes or cyclones and various other weather elements can affect the yield of a given year or season. Having diverse yields allows the farmer to have some income even if the primary crop gets damaged or doesn't yield as much as expected.

Optimum use of Soil

Intercropping makes the most of the available soil. When anything is grown on a farm-land, the crop tends to absorb as much water and nutrients as it needs. There could be more nutrients in the soil under the crops and around. This soil and more specifically the nutrients can be used, by the different varieties of crops. Intercropping also averts soil runoff and can prevent the growth of weeds.

Good for Primary Crops

Intercropping is good for the primary crops. The secondary crops can provide shelter and even protect the primary crops. Intercropping also allows you to grow cash crops or any crop that will actually supplement the primary crop in some way.

List of Disadvantages of Intercropping

Poor Yields

Intercropping can lead to poorer yields. The crops may not be compatible. The crops may actually compete for the same nourishment and of course water, which may lead to an unmanageable conflict. It is possible that both crops don't yield enough produce.

Costly and Complicated

Intercropping obviously costs more money upfront. There is a need for more fertilizers and water. Harvesting is also more complicated. If something goes horribly wrong with either crop then the other crop may also get damaged.

Monocropping

Monocropping is the agricultural practice of growing a single crop year after year on the same land, in the absence of rotation through other crops or growing multiple crops on the same land polyculture. Corn, soybeans, and wheat are three common crops often grown using monocropping techniques.

Monocropping allows for farmers to have consistent crops throughout their entire farm. They can use the same seed, pest control, machinery and growing method on their entire farm giving them a larger yield for a significantly lower cost.

Soil Ecology

While economically a very efficient system, allowing for specialization in equipment and crop production, monocropping is also controversial, as it damages the soil ecology (including depletion or reduction in diversity of soil nutrients) and provide an unbuffered niche for parasitic species, increasing crop vulnerability to opportunistic insects, plants, and microorganisms. The result is a more fragile ecosystem with an increased dependency on pesticides and artificial fertilizers. The concentrated presence of a single cultivar, genetically adapted with a single resistance strategy, presents a situation in which an entire crop can be wiped out very quickly by a single opportunistic species. An example of this would be the potato famine of Ireland in 1845–1849, and according to Devlin Kuyek is the main cause of the current food crisis with monoculture rice crops failing as the effects of climate change become more acute.

Strategy

Monocropping as an agricultural strategy tends to emphasize the use of expensive specialized farm equipment—an important component in realizing its efficiency goals. This can lead to an increased dependency on fossil fuels and reliance on expensive machinery that cannot be produced locally and may need to be financed. This can make a significant change in the economics of farming in regions that are accustomed to self-sufficiency in agricultural production. In addition, political complications may ensue when these dependencies extend across national boundaries.

The controversies surrounding monocropping are complex, but traditionally the core issues concern the balance between its advantages in increasing short-term food production—especially in hunger-prone regions—and its disadvantages with respect to long-term land stewardship and the fostering of local economic independence and ecological sustainability. Advocates of monocropping tend to claim that in its absence many human populations would be reduced to starvation or to a degraded level of civilization comparable to the Dark Ages. On the other hand, critics of monocropping dispute these claims and attribute them to corporate special interest groups, citing the damage that monocropping causes to societies and the environment.

Monocropping Difficulty

A difficulty with monocropping is that the solution to one problem—whether economic, environmental or political—may result in a cascade of other problems. For example, a well-known concern is pesticides and fertilizers seeping into surrounding soil and groundwater from extensive monocropped acreage in the U.S. and abroad. This issue, especially with respect to the pesticide DDT, played an important role in focusing public attention on ecology and pollution issues during the 1960s.

Soil depletion is also a negative effect of monocropping. Crop rotation plays an important role in replenishing soil nutrients, especially atmospheric nitrogen converted to usable forms by nitrogen-fixing plants used in fallow fields. In addition, it performs an important role in preventing pathogen and pest build-up. In a monocropping regime, farmers are less likely to rotate their crops and replenish such essential soil nutrients. In addition, artificial high-nitrogen fertilizers can "burn" the soil by creating an unfavorable environment for indigenous organisms, a phenomenon well-known to organic gardeners and farmers (who avoid it), resulting in further disruption of soil ecology and dependence on further short-term fertilizer strategies. Lacking a stable ecology, in the absence of substantial irrigation and chemical "fixes", the soil can become dry and begin to erode. As the soil becomes arid and useless, the need for more land becomes an issue, leading to the destruction of even more land—a high-tech version of slash and burn agriculture.

Deforestation

Under certain circumstances monocropping can lead to deforestation or the displacement of indigenous peoples. For example, since 1970 the Amazon Rainforest has lost nearly 1/5 of its forest cover. A main cause of this deforestation is local farmers clearing land for more crops. In Colombia, the need for more farming land is causing the displacement of large populations of peasants.

Subsidies

In order to help reduce dependence on fossil fuels the U.S. government subsidizes the monocropping of corn and soybeans to be used in ethanol production. However, monocropping itself is highly chemical- and energy-intensive, as studies by Nelson (2006) indicate. Such studies have shown that the "hidden" energy costs associated with producing each unit of bio-fuel are significantly larger than the amount of energy available from the fuel itself.

Advantages and Disadvantages of Monocropping

Advantages of Monocropping

- This approach to farming is fairly simple in nature, focusing all its needs and preference on one single crop species. Farmers just need to prepare the soil, and irrigate the land based on one crop. With monoculture, the field is in a better position to provide maximum output for a particular crop.

- Harvesting becomes fairly easy as the desired parts of the plant can be easily assembled without damaging other plants, which would be very difficult in polyculture. Chemical treatment is feasible, pests and diseases can be treated without having to worry about their side effects on other plants.

- It helps to keep down farming costs down. Farmers yield more output in less resources. Makes management pretty easy, machines and various methods can be utilized more efficiently and systematically.

- The knowledge of single plant species is sufficient for a good crop, farmers need not worry about other species, their cultivation methods, disease prevention, etc. Since the emphasis is on one plant, acquiring adequate knowledge or expertise is also easy.

- It is convenient for home gardeners who want to have a bigger harvest of a particular plant; suppose they want to save up on corn or barley to reduce their expenses. Growing a single large crop requires less investment.

- By grouping different plants together, farmers or gardeners have to cater to the fertilizer requirement of different plants. But with monoculture, they can easily use and apply one common fertilizer for all the plants.

- Planting same species of crop is much easier and faster process. Farmers can prepare garden beds and seed plants altogether. They just need to prepare garden beds for only one type of crop.

- Controlling pests and disease becomes relatively easy. Growers just need to use one pesticide for all the plants, because the diseases affecting them would be common.

- There is less competition for sunlight, nutrients, and space from other species. It helps to control other undesirable growth. It helps to maximize profits by planting crops which yields high gross margin.

- High gross margin crops are market-driven, and it's easy to market such crops. Farmers particularly plants crop which can be consumed all year round, and also those which will thrive under all weather conditions.

Disadvantages of Monocropping

- Monoculture does not support other flora and fauna. According to its definition, other plants should not be planted. We all need different environment to survive; likewise, animals continuously living in one environment will lack the feel of a natural habitat.

- If a particular disease or pest can affect one single plant, then it can possible affect all the other plants as they also will be vulnerable to their attack. An infected plant, in this scenario, will be surrounded by infected plants, which will lead to the destruction of the entire crop.

- Plants require multiple resources to thrive; however, if a crop is planted in the same field for extended periods, it limits its chance of taking advantage of other nutrients in the soil.

- One of the problems of monoculture farming is limited food options. For the sake of saving their resources, farmers plant one single crop, leaving consumers with few options to survive on, which can lead to malnutrition, especially in developing countries.

- Due to the cultivation of same crops over and over again, monoculture reduces the nitrogen composition in the soil. Once the land is used for one single crop, soil fertility diminishes at a faster rate.

- Because of diminishing soil fertility, farmers rely heavily on chemicals and technology to promote plant growth and production. Monoculture leads to environmental damage when the chemicals and pesticides make their way into ground water.

- Due to major crop failure, farmers can suffer high losses, which in turn would contribute to total market loss. Farmers depend on one type of production, so their income is also not stable.

- Monoculture results in less diversity of other species, this applies to both plants and animals. This, in turn is not good for the bio-diversity of that entire region.

- Monoculture is not advocated because repetitive use of fertilizers can lead to soil erosion, which makes it difficult for plants to grow.

- Planting crops over a large area can be time-consuming for a farmer. Not to mention the efforts and investment required to set up a complex irrigation system.

References

- Intercropping-principles: doccentre.info, Retrieved 26 August, 2019

- 6-advantages-and-disadvantages-of-intercropping: connectusfund.org, Retrieved 3 March, 2019

- Victor m. Quintana s. (august 29, 2007). "biofuels and small farmers". Translated by annette ramos. Archived from the original on december 10, 2008. Retrieved december 6, 2008

- Advantages-types-and-disadvantages-of-multiple-cropping-system: 1001artificialplants.com, Retrieved 13 May, 2019

- "Deforestation in the amazon – council on foreign relations". Council on foreign relations. Retrieved 2018-04-12

- Advantages-disadvantages-of-monoculture: gardenerdy.com, Retrieved 10 January, 2019

3

Crop Diseases

Crop diseases are often caused by pathogens such as fungi and virus. Gibberella fuji-kuroi, Athelia rolfsii, false smut, sheath rot, helminthosporium oryzae, bacterial leaf blight are the most common diseases that affect crops. This chapter has been careful-ly written to provide an easy understanding of these different types of crop diseases.

Crop diseases are as old as agriculture itself and there are many examples in history where people have suffered as a consequence of plant diseases. For example, the Ro-mans used to sacrifice a red dog every year to appease the rust goddess Rubiga. Blight of potato caused the Great Famine in Ireland. More recently, Ascochyta blight decimat-ed Wimmera chickpea crops in the late 1990s and brought hardship to many farmers.

Diseases cause economic losses to crops by attacking the plant at many different places and stages in its growing cycle. Some will cause damage by interfering with water and mineral absorption from the soil (diseases of the roots and stem base), some will affect photosynthesis by killing the leaves of the plant (diseases of the foliage), some will im-pair translocation of sugars produced in the photosynthesis to the grain (systemic virus diseases) and yet others will completely destroy developing grains (disease of the head and kernel).

Diseases remain a constraint to field crop production. They are estimated to causes annual losses valued at around $120 million for Victoria. Without on-going, current control practices this loss could be much higher.

Fungal Diseases

Sphacelotheca Reiliana

Sphacelotheca reiliana, also known as Sporisorium reilianum, is a species of biotrophic fungus in the family Microbotryaceae. It is a plant pathogen that infects maize and sorghum.

Hosts and Symptoms

Sporisorium reilianum causes the diseases maize head smut and sorghum head smut. This soil borne smut fungus has two formae speciales. S. reilianum f. sp. reilianum is specific to sorghum and S. reilianum f. sp. zeae is specific to maize. It is unknown why the two formae speciales cannot form spores on their respective non-favored hosts.

Symptoms of the fungus are expressed on both the tassels of corn and sorghum as well as on the actual ear in the form of large smut galls. When the sorghum tassel is infected, the fragile gall membrane will have a range from just a few black spores to a large mass of black spores covering the tassel. The spores are a sign of the disease and are used for dispersal of the disease to other corn and sorghum plants. When the ear of the corn is infected, it looks very small and tear-drop shaped and seems as though it does not have a cob inside at all. The cob is replaced by white sori which are the structures that make and hold the spores of the fungus. If there is an infected tassel, it is likely that the ears will also have head smut.

Life Cycle

Sporisorium reilianum is noted to have a sexual stage in its disease cycle similar to that of *Ustilago maydis*. Initial infections occur on roots of young seedlings. The pathogen develops systemically and is found on ear and tassel tissues as the host plant matures. At maturity teliospores can be found in the white sori of the infected heads of corn. These will be easily dispersed by the wind. Favorable nutritive soil and weather conditions around 23-30 °C allows for germination of the teliospores in the soil. Generation of a four-celled basidium occurs, leading to haploid basidiospores that create sporidia. These sporidia fuse due to a compatibility or likeness that induces the formation of dikaryotic mycelium, which is infectious and parasitic. This intracellular mycelium can be found invaded in parts of the flowering development of the corn, and *S. reilianum* can completely decrease floral tissue due to an ability to detect floral induction.

S. reilianum is biotrophic in that it depends on the maize or sorghum for growth and survival. The inflorescence of the male or female parts of the plants, female being the ear and the male being the tassel can be affected by the timing of infection by this species. Necrosis and disease development is most prevalent on the head of the infected host.

Environment

Maize head smut occurs in most maize-growing areas, including many regions of North America, Australia, Asia, and southern Europe. It causes tremendous loss of yield during outbreaks due to the replacement of the ear with large smut sori. To infect maize roots, *S. reilianum* f. sp. *zeae* must form a dikaryotic parasitic hypha which results for the mating of two compatible haploid strains. The infection always occurs in soil via the root, unlike *Ustilago maydis*, another maize smut, which infects maize plants via aerial parts. Temperatures of 23-30 °C are optimal for field infection of corn, suspected to be due to maximum teliospore germination.

Management

Detection of *S. reilianum* is a key step toward the development of an effective disease management system. Conventional methods, including pathogen isolation and microscopic morphological study, are labor-intensive and cumbersome, and sometimes yield inconclusive results. Rapid, specific and sensitive molecular tools, such as PCR, are required for the detection and evaluation of *S. reilianum*. In 1999 there was developed a DNA-based assay for the detection of *S. reilianum* and its differentiation from *Ustilago maydis*. More recently, in 2012, a method using Fourier transform infrared spectrometry was established for identification of *S. reilianum* spores.

Head smut can be controlled by use of resistant cultivars and seed treatment with fungicides. Foliar application of fungicides have not controlled the disease. Crop rotation is of questionable value because the smut spores can survive for long periods in the soil. It has been suggested that the most economical and environmentally friendly method to reduce maize yield losses is to breed and deploy resistant maize hybrids. In comparison with other cereal crops, such as rice (*Oryza sativa*) and wheat (*Triticum aestivum*), maize has fewer qualitative resistance genes that have been used extensively by breeders. Instead, maize has relatively more quantitative resistance loci deployed in the field to counter the majority of diseases. In other words, maize resistance to *S. reilianum* is thought to be under polygenic control. The cumulative effects of multiple smaller-effect quantitative resistance loci can produce high or even complete resistance.

Importance

The pathogenic tendency of *Sporisorium reilianum* as illustrated above is to carry out its life cycle in congruence with the growth of either maize or sorghum as the host plant. For example, In place of ears of corn; plant-like sorus develop. The onset of this smut is sporadic and slowly progressive however disease severity is high. For example, a field of corn can have 10% infected with yield loss that has an 80% apparent infection rate. Post infection treatment to eliminate or reduce the disease are marginally effective or ineffective. The massive crop loss that this pathogen can create is devastating to producers who are unaware of the teliospores overwintering in the soil; which, can survive for many years. Treatment of seeds with a fungicide during planting or prior to harvest is important to limit the spread of spores especially if high risk field areas known by the producer. It is important to limit transmission of spores to disease free areas via harvesting equipment or planting tools. The incidence of *Sporisorium reilianum* occurs in occurs in Africa, Europe, U.S., and China. Fungicide use in the market today indicates that there are about 52 million acres of cereals lost worldwide a year.

Pathogenesis

Sporosorium relianium has shown the ability to sense the presence of nearby plants. This pathogen always infects via the roots, and hyphal proliferation near the roots is a characterization of the early stage of this Fungal pathogen. During the Basido-phase, basidiospores grow as haploid saprophytic yeasts. These yeasts join to form dikaryotic hyphae which in turn infect the host plant via the roots. It has been demonstrated on two varieties of maize that plant root exudates impact the growth and branching of *S. relianium* endomycorrhizal fungi prior to fungal contact with the root.

Sorghum Downy Mildew

Sorghum downy mildew (SDM) is a highly destructive disease caused by the soilborne fungus Peronosclerospora sorghi. It can have a significant economic impact, as infection results in plant death or lack of grain formation in the panicle. It is most prevalent in the coastal bend and southern regions of Texas.

Infection/Life Cycle

Grain sorghum plants can be infected in one of two ways: systemically or locally. Systemic infection occurs by overwintering fungal bodies found in the soil or on residue known as oopsores. The oopsores infect the plant as they move with moisture and are taken up through the roots. The plant will display symptoms of a systemic infection at three to four weeks after planting. As the disease progresses leaf tissue shreds and mature oopsores are released and fall to the soil and the cycle can repeat in the next year.

Localized infection occurs from airborne spores, known as conidia, which are produced in cool and humid conditions. The conidia appear as a white, fuzzy growth on the underside of leaves. The spores are carried by the wind and will infect leaves of neighboring plants. Local infection can be widespread throughout the field, but is usually short-lived and causes no yield loss. New infections cease as temperatures increase; however, under cool and wet conditions conidia can cause systemic infection in young seedlings.

Symptoms

- Systemically infected seedlings will have a pale-yellow to light-colored streaked.

- Streaking generally occurs from the lower leaves, moving upward as the disease progresses.

- Later-emerging leaves have parallel stripes of green and white tissue.

- Striped areas eventually turn necrotic and have a shredded appearance as the plant grows larger.

- Systemically infected plants will not produce any grain, resulting in a barren plant.

- In cool, humid conditions a fuzzy, white growth ("downy" growth) will occur on the underside of the infected leaf tissue.

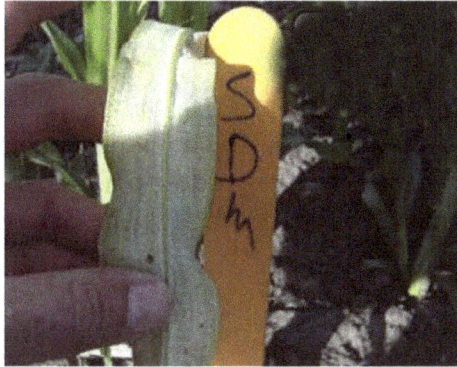

- Local infection is caused by spores from the "downy" growth being blown onto healthy leaf tissue; these lesions are generally brown-colored and blocky in appearance.

Control

Genetic resistance is the first line of defense in SDM control, and can be further complemented with the use of a fungicide seed treatment. Pioneer continues to be the industry leader in developing grain sorghum hybrids with genetic resistance to the various pathotypes of SDM. Additionally, all Pioneer brand sorghum seed comes treated with a proven combination of Maxim (fludioxinil) and Apron (mefanoxam-m) fungicides, which not only provide control against SDM but other seedling diseases as well. Crop rotation is also necessary as SDM pathotypes can mutate when exposed to a viable host year after year. Eliminating alternate hosts such as Johnsongrass and shattercane, even during crop rotation can help to control disease levels in subsequent cropping years.

Rust (Fungus)

Rusts are plant diseases caused by pathogenic fungi of the order Pucciniales (previously also known as Uredinales).

An estimated 168 rust genera and approximately 7,000 species, more than half of which belong to the genus *Puccinia*, are currently accepted. Rust fungi are highly specialized plant pathogens with several unique features. Taken as a group, rust fungi are diverse and affect many kinds of plants. However, each species has a very narrow range of hosts and cannot be transmitted to non-host plants. In addition, most rust fungi cannot be grown easily in pure culture.

A single species of rust fungi may be able to infect two different plant hosts in different stages of its life cycle, and may produce up to five morphologically and cytologically distinct spore-producing structures viz., spermogonia, aecia, uredinia, telia, and basidia in successive stages of reproduction. Each spore type is very host specific, and can typically infect only one kind of plant.

Rust fungi are obligate plant pathogens that only infect living plants. Infections begin when a spore lands on the plant surface, germinates, and invades its host. Infection is limited to plant parts such as leaves, petioles, tender shoots, stem, fruits, etc. Plants with severe rust infection may appear stunted, chlorotic (yellowed), or may display signs of infection such as rust fruiting bodies. Rust fungi grow intracellularly, and make spore-producing fruiting bodies within or, more often, on the surfaces of affected plant parts. Some rust species form perennial systemic infections that may cause plant deformities such as growth retardation, witch's broom, stem canker, galls, or hypertrophy of affected plant parts.

Rusts get their name because they are most commonly observed as deposits of powdery rust-coloured or brown spores on plant surfaces. The Roman agricultural festival Robigalia has ancient origins in combating wheat rust.

Impacts

Rusts are considered among the most harmful pathogens to agriculture, horticulture and forestry. Rust fungi are major concerns and limiting factors for successful cultivation of agricultural and forestry crops. White pine blister rust, wheat stem rust, soybean and coffee rust are examples of notoriously damaging, economically important crops.

Life Cycle

Germinating spore of Puccinia graminis, model from the late
19th century, Botanical Museum Greifswald.

All rusts are obligate parasites, meaning that they require a living host to complete their life cycle. They generally do not kill the host plant but can severely reduce growth and yield. Cereal crops can be devastated in one season and trees that get infected in the main stem within their first five years by the rust Cronartium quercuum often die.

Rusts can produce up to five spore types from corresponding fruiting body types during their life cycle, depending on the species. Roman numerals have traditionally been used to refer to these morphological types.

- Pycniospores (Spermatia) from Pycnidia. These serve mainly as haploid gametes in heterothallic rusts.

- Aeciospores from Aecia. These serve mainly as non-repeating, dikaryotic, asexual spores, and go on to infect the primary host.

- Urediniospores from Uredia (Uredinia). These serve as repeating dikaryotic vegetative spores. These spores are referred to as the repeating stage because they can cause auto-infection on the primary host, re-infecting the same host from which the spores were produced. They are often profuse, red/orange, and a prominent sign of rust disease.

- Teliospores from Telia. These dikaryotic spores are often the survival/overwintering stage of life cycle. Later they germinate to produce basidia.

- Basidiospores from Teliospores. These haploid spores often infect the alternate host in Spring. although these are rarely observed outside of the laboratory.

Rust fungi are often categorized by their life cycle. Three basic types of life cycles are recognized based on the number of spore states as macrocyclic, demicyclic, and microcyclic. The macrocyclic life cycle has all spore states, the demicyclic lacks the uredinial state, and the microcyclic cycle lacks the basidial, pycnial, and the aecial states, thus possess only uredinia and telia. Spermagonia may be absent from each type but especially the microcyclic life cycle. In macrocyclic and demicyclic life cycles, the rust may be either host alternating (heteroecious), i.e., the aecial state is on one kind of plant but the telial state on a different and unrelated plant, or non-host alternating (autoecious), i.e., the aecial and telial states on the same kind of plant. Heteroecious rust fungi require two unrelated hosts to complete their life cycle, with the primary host being infected by aeciospores and the alternate host being infected with basidiospores. This can be contrasted with an autoecious fungus which can complete its life cycle on a single host species. Understanding the life cycles of rust fungi allows for proper disease management.

Host Plant-rust Fungus Relationship

There are definite patterns of relationship with host plant groups and the rust fungi that parasitize them. Some genera of rust fungi, especially *Puccinia* and *Uromyces*, comprise species that are capable of parasitizing plants of many families. Other rust genera appear to be restricted to certain plant groups. Host restriction may, in heteroe-cious species, apply to both phases of life cycle or to only one phase.

Infection Process

The fungi produce asexual spores which disperse by wind, water or by insect vectors spreading the infection.

Rust fungi are biotrophs, taking nutrients from living cells. When airborne spores settle on a plant, weak hydrophobic interactions are formed with the cutin on the plant cell surface, securing it. By a process not fully understood, the production of mucous like substances called 'adhesins', initially stick the spore to the plant surface.

Once attached, the spore germinates by growing a germ tube and eventually locates a stoma by a touch responsive process known as thigmotropism. This involves growing towards a ridge between the epidermal cells, followed by a perpendicular growth which end at the stoma.

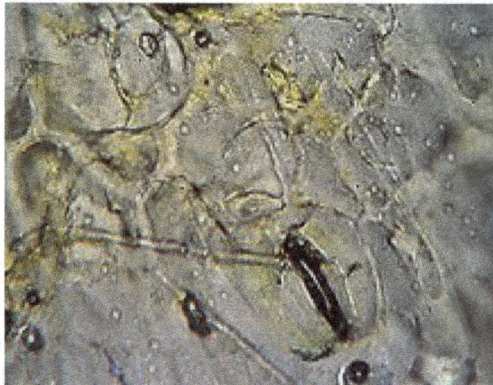

Rust hypha attacking stoma (1600x magnification).

Inside the stoma, the hyphae tips flatten out to form structures known as appressoria that lock to the cell walls. It is thought that the whole process is mediated by a mech-ano-sensitive calcium ion channel, located within the germ tube tip, which produces electric currents that stretch the cell membranes, changing gene expression and form-ing the appresorium.

Then a peg grows into the plant's mesophyll cells. The peg produces specialised hyphal tips, known as haustoria. These spread around the plant cells without invading the membranes. The plant cell membranes invaginate around the main haustorial body forming a space known as the extra-haustorial matrix. An iron and phosphorus rich neck band bridges the plant and fungal membranes in the space between the cells for water flow, known as the apoplast, thus preventing the nutrients reaching the plant's cells. The haustorium contains amino acid and hexose sugar transporters and H^+-ATPases which are used for active transport of nutrients from the plant nourishing the fungus. It continues growing until spore growth occurs. The process repeats every 10 – 14 days, producing numerous spores, carried by wind to new hosts.

Common Rust Fungi in Agriculture

- Cronartium ribicola (White pine blister rust); the primary host are currants, and white pines the secondary. Heterocyclic and macrocyclic,

- Gymnosporangium juniperi-virginianae (Cedar-apple rust); Juniperus virginiana is the primary (telial) host and apple, pear or hawthorn is the secondary (aecial) host. Heteroecious and demicyclic,

- Hemileia vastatrix (Coffee rust); Primary host is coffee plant; unknown alternate host. Heteroecious,

- Phakopsora meibomiae and P. pachyrhizi (Soybean rust); Primary host is soybean and various legumes. Unknown alternate host. Heteroecious,

- Puccinia coronata (Crown Rust of Oats and Ryegrass); Oats are the primary host; Rhamnus spp. (Buckthorn) is alternate host. Heteroecious and macrocyclic,

- Puccinia graminis (Stem rust of wheat and Kentucky bluegrass, or black rust of cereals); Primary hosts include: Kentucky bluegrass, barley, and wheat; Common barberry is the alternate host. Heteroecious and macrocyclic,

- Puccinia hemerocallidis (Daylily rust); Daylily is primary host; Patrina sp is alternate host. Heteroecious and macrocyclic,

- Puccinia persistens subsp. triticina causes wheat rust in grains. It is also known as 'brown or red rust',

- Puccinia sorghi causes common rust in corn,

- Puccinia striiformis causes 'Yellow rust' in cereals,

- Uromyces appendiculatus causes Bean rust in common bean (Phaseolus vulgaris),

- Uromyces phaseoli ; Primary host: bean. Autoecious and macrocyclic,

- Puccinia melanocephala causes 'Brown rust' in sugarcane,

- Puccinia kuehnii causes 'Orange rust' in sugarcane.

Management of Rust Fungi Diseases

The control methods of rust fungus diseases depend largely on the life cycle of the particular pathogen. The following are examples of disease management plans used to control macrocyclic and demicyclic diseases:

1. Macrocyclic Disease: Developing a management plan for this type of disease depends largely on whether the repeating stage (urediniospores) occur on the economically important host plant or the alternate host. For example, the repeating stage in white pine blister rust disease does not occur on white pines but on the alternate host, Ribes spp. During August and September Ribes spp. give rise to teliospores which infect white pines. Removal of the alternate host disrupts the life cycle of the rust fungi Cronartium ribicola, preventing the formation of basidiospores which infect the primary host. Although spores from white pines cannot infect other white pines, survival spores may overwinter on infected pines and reinfect Ribes spp. the following season. Infected tissue is removed from white pines and strict quarantines of Ribes spp. are maintained in high risk areas.

Puccinia graminis is a macrocyclic heteroecious fungus that causes wheat stem rust disease. The repeating stage in this fungus occurs on wheat and not the alternate host, barberry. The repeating stage allows the disease to persist in wheat even though the alternate host may be removed. Planting resistant crops is the ideal form of disease prevention, however, mutations can give rise to new strains of fungi that can overcome plant resistance. Although the disease cannot be stopped by removal of the alternate host, the life cycle is disrupted and the rate of mutation is decreased because of reduced genetic recombination. This allows resistance bred crops to remain effective for a longer period of time.

2. Demicyclic Disease: Because there is no repeating stage in the life cycle of demicyclic fungi, removal of the primary or the alternate host will disrupt the disease cycle. This method, however, is not highly effective in managing all demicyclic diseases. Cedar-apple rust disease, for example, can persist despite removal of one of the hosts since spores can be disseminated from long distances. The severity of Cedar-apple rust disease can be managed by removal of basidiospore producing galls from junipers or the application of protective fungicides to junipers.

Domestic Control

Rust is very hard to treat. Fungicides such as Mancozeb or Triforine may help but may never eradicate the disease. Some organic preventative solutions are available and sulphur powder is known to stop germination. High standards of hygiene and good soil drainage and careful watering may minimise problems. Any appearance of rust must be immediately dealt with by removing and burning all affected leaves. Composting, or leaving infected vegetation on the ground will spread the disease.

Commercial Control

In large plantations in USA, fungicides are applied by air. The process is expensive and fungicide application is best reserved for seasons when foliar diseases are severe. Research indicates, the higher the foliar disease severity, the greater the return from the use of fungicides. There are a variety of preventative methods that can also be employed.

- Symptoms of rust disease are correlated to relatively high moisture. The avoidance of overhead watering at night, using drip irrigation, reducing crop density, and using fans to circulate air flow will lower the relative moisture and decrease the severity of rust infection,

- The use of rust resistant plants,

- Crop rotation can break the disease cycle because many rusts are host specific,

- Inspect all imported plants and cuttings for symptoms. It is important to continuously observe these plants because rust diseases have a latent period (plant has the disease but shows no symptoms),

- Many crops, such as wheat, are replanted with disease-free seed.

Loose Smut

Loose smut of barley is caused by *Ustilago nuda*. It is a disease that can destroy a large proportion of a barley crop. Loose smut replaces grain heads with smut, or masses of spores which infect the open flowers of healthy plants and grow into the seed, without showing any symptoms. Seeds appear healthy and only when they reach maturity the following season is it clear that they were infected. Systemic fungicides are the major control method for loose smut.

Hosts and Symptoms

The major symptom of loose smut is the "smutted" grain heads, which contain masses of black or brown spores where the grain would normally be. The spores completely replace the grain head so that there is no grain to be harvested on infected plants. It may be possible to identify infected plants in the field before they reach the flowering stage by looking for plants which are taller and more mature than the rest of the field. The fungus causes infected plants to grow slightly taller and mature slightly sooner than the uninfected plants in the field. Since it must infect through the open florets, this gives the fungus a competitive advantage by allowing it to fall down to the healthy plants and ensuring that the fungus has a little extra time to produce and disperse spores before the florets of the healthy plants open.

Ustilago nuda infects barley '*Hordeum vulgare* L.' but there are many strains of *Ustilago* which infect many different cereal crops in a similar manner. *Ustilago tritici*, for example, is loose smut of wheat.

Disease Cycle

The disease cycle of loose smut begins when teliospores are blown to open flowers and infect the ovary either through the stigma or directly through the ovary wall. There are multiple mating types for Ustilago spp. so infection will only occur if two compatible mating types are present in the same flower.

After landing in an open floret, the teliospores give rise to basidiospores. Without dispersing to any alternate host plant, the basidiospores germinate right where they are. The hyphae of two compatible basidiospores then fuse to establish a dikarytic stage.

After germination inside the ovary, the fungal mycelia invade the developing embryo in the seed. The fungus stays alive in the seed until the next growing season, when it is planted along with the seed. As the developing plant grows, the fungus grows with it. Once it's time for the flowers to form, teliospores are produced in place of the flowers and develop where the grain would be.

Plants which are infected with Ustilago spp. actually grow taller and flower earlier than their healthy counterparts. This gives the infected plants an advantage in that the flowers of uninfected plants are more physically and morphologically susceptible to infection. The teliospores in the smutted grain heads disperse to the open flowers of the healthy plants, and the cycle continues.

Environment

Loose smut is unique in that it needs to infect the seed in the previous growing season in order to be a pathogen of the plant in the current growing season. This means that regardless of environmental conditions, if an infected seed is planted as long as the growing conditions are good for the plant, they will be good for the fungus. However, if there are environmental stressors on the plant such as drought or extreme heat, the plant will be more susceptible to injury because of the added stress from the fungal infection. If the plant doesn't survive to its reproductive stage, the fungus will not get to reproduce either.

Environment becomes a factor once teliospores have been produced and are ready for dispersal. In order to get to the open flowers of healthy plants, the teliospores must be moved by wind or rain or possibly insects. If dry, calm conditions persist for the entire time the flowers are open, the infection rate will be low. Wind and moderate rain, as well as cool temperatures (16–22 °C) are ideal for the dispersal of spores. Conditions in Eastern Europe, Western Siberia, and the Northern Caucasus regions are some of the most conducive to loose smut.

Control

The most widely used method of control for loose smut is using treated seed. Seeds are typically treated with a systemic fungicide to kill any fungus that may be inside. Since the fungus grows up with the plant, it's very important that the fungicide used to treat the seed be systemic and not just external. The most common type of systemic fungicides used for loose smut are from the Carboxin group of chemicals. Although carboxins are highly toxic to fish and mildly toxic to other animals, they degrade rapidly in the environment and since there the fungicide is coated on seed and not sprayed there is not much loss to the environment.

Barley varieties with resistance to loose smut are also commercially available. Although technology is being researched and used to help speed the process of resistant variety development, these traditional breeding methods are still very slow, and it is difficult to develop varieties with resistance which also possess other desirable traits such as those for yield and grain quality. Another option is heat treating the seed to kill the fungus before planting, but this is a delicate process because too much heat will kill the plant embryo and not enough will allow the fungus to survive. Research has also been done to use radio frequency heating to treat seed.

Importance

If uncontrolled, loose smut can wipe out entire crops, since it replaces the grain. In areas where people depend on their grain crops for survival and don't have the money or technology resources to control it, the disease can be devastating. Not only does it wipe out the crop, growers cannot even try again next year since any seeds they were able to harvest will be infected and will not produce seed the following season.

Loose smut has recently become a more serious problem in the Ladakh region in the Himalayan mountains of India. This area has been particularly affected by global climate change and has experienced many new crop pests and diseases because of it. Barley is an important staple crop in this region and although loose smut is not the most prevalent disease of barley in the region, it has the potential to develop into a more serious problem in coming years.

In places like the United States where there are ample resources and technologies such as fungicide seed treatment, loose smut is not an especially important disease.

Postbloom Fruit Drop

Postbloom fruit drop (PFD) of citrus is caused by the fungus, Colletotrichum acutatum J. H. Simmonds. This fungus infects citrus flowers producing orange-brown necrotic spots or affecting the entire petal. Fruitlets on affected inflorescenses do not develop or abscise, but the calyx and floral discs remain attached to the peduncle. These persistent calyces, commonly called buttons, are diagnostic for the disease and may persist for many months after flowering.

Necrotic spots on petals of sweet orange caused by Colletotrichum acutatum.

Persistent calyces produced following flower infection and abscission of fruitlets.

PFD can be severe when the flowering period is extended and rains occur during the bloom. Losses may approach 100% in some locations in certain years. Yield loss to PFD in Florida is sporadic with serious epidemics in some years and virtually no losses in others. In São Paulo State, Brazil, epidemics are more frequent and some losses may occur annually. In humid, tropical areas such as Belize, southern Mexico, and Costa Rica, PFD is a limiting factor in citrus production.

Properly timed applications of fungicides, such as benomyl, have been effective in reducing losses to PFD. A model was developed in Florida to predict epidemics and to schedule fungicide applications. The model was based originally on the number of infected flowers observed on 20 trees and the amount of rainfall during the last 5 days. It was subsequently modified to include the leaf wetness duration following rainfall events. Fungicide applications timed according to the model increased fruit counts 30 to 600% and use of the model prevented unnecessary fungicide applications.

However, even in Florida, the model has some shortcomings. In cases where some infected flowers are present and high rainfall (> 25 mm) associated with cold fronts occurs, an application of fungicide is usually indicated by the model. However, in such events, moisture dries quickly, the rain is followed by cool temperatures, and PFD epidemics rarely develop. Inclusion of leaf wetness duration in the model helped minimize that problem, but the model is additive and even if the leaf wetness factor is zero, sprays are often indicated. Thus, the model is conservative and occasionally indicates sprays that are not necessary.

The model as developed in Florida cannot be used effectively in other locations. The fungus survives periods without flowers as appressoria on leaf surfaces. The model assumes that inoculum must be dispersed by rains from localized sources. However, in tropical areas, citrus trees often flower 2 to 3 times per year and PFD occurs every year. Thus, inoculum is present in large amounts in virtually every tree. Multiple rains are not needed to promote inoculum build up and 1 to 2 rain events or even heavy dews or fogs can result in considerable.

The PFD model continues to be very useful in Florida and is often effective in subtropical areas such as southern São Paulo State in Brazil, but a more widely applicable system would be desirable. The decision support system described herein includes many of the principles developed in the PFD models, but also incorporates the experience of the authors in working with PFD and its control in Brazil, southern Mexico, Central America, and the Caribbean.

Development of the PFD-FAD System

Many factors are considered in the assessment of the risk of PFD and the need for fungicide applications. First, we must consider risk factors that affect PFD incidence inherent in any planting. Climatological factors can affect PFD directly, but can also have indirect impacts. Locations or species of citrus with multiple blooms are more prone to disease because inoculum may be maintained at high levels all year. Some species or cultivars of citrus are more susceptible than others. The presence of declining trees which flower out of season may increase PFD incidence especially in areas with a single annual bloom. In addition, we must consider factors which could affect inoculum build-up such as the frequency of occurrence of PFD in previous years, the number of buttons remaining from the previous year, and occurrence of PFD on scattered flowers appearing prior to the main bloom. The number of rainfall events that occurred during the main bloom period but prior to the period under consideration are also taken into account.

The primary factors determining the potential for an epidemic in the immediate future are inoculum availability, rainfall, and leaf wetness duration, and these are basically the factors considered in the PFD model. However, the PFD-FAD system provides more flexibility in dealing with these factors than the PFD model did. Lastly, the stage and amount of bloom present is critical in decisions as to whether or not to make fungicide applications.

1. Stage of bloom: This factor is considered first because if the situation at flowering time is not appropriate, no sprays are indicated regardless of the history, inoculum presence, or weather factors.

2. Sufficient flowers: In order to justify an application, there must be sufficient flowers in the planting to pay for the application if fruit set is increased. Research in Florida indicates that for each 100 buttons formed, 6 fruit are lost. However, the percentage of flowers which set fruit varies tremendously in different growing areas. Also, the value of fruit varies greatly with the species or cultivar involved, the time of year that the fruit will be harvested, and the economic conditions and costs of applications.

In many areas, losses to PFD can be substantial and orchard care costs must be paid regardless of the yield or losses to PFD. Thus, fungicide applications may be economically justified even when fruit prices are low. Fungicide sprays may be needed in order to have some yield to offset care costs even if net return is negative. Fungicide applications are not justified only if the crop will have little value and it is likely that the fruit will not be harvested.

In the more temperate citrus-growing areas with one bloom per year, flowering is abundant and occurs once per year. In these situations, decisions are less complicated. However, in tropical areas flowering often occurs 2 to 3 times per year and may vary greatly in intensity within and between groves. Poorly-managed plantings often have weak, sporadic blooms and there are insufficient flowers to justify an application. Thus, it is often more important to improve fertilization, weed control, water management, and control of greasy spot than to invest in fungicides for PFD control. Uniform blocks of well-foliated vigorous trees are much more likely to produce the abundant and uniform blooms needed to justify fungicide applications for PFD. Ultimately, the grower must make the economic decision as to whether it is worthwhile to even consider applications for control of PFD.

3. Inoculum availability: The other major consideration in assessing the need for fungicide applications is the effect of bloom stage on inoculum availability. Applications made early in the bloom, prior to the presence of open flowers on the main bloom are largely ineffective. Benomyl applied to diseased flowers does not kill spores nor reduce the potential for future disease development. If no rain occurs or few open flowers are available, these affected petals simply fall to the ground where they do not represent a potential for spread. While flower buds and unopened flowers can be infected, these flowers are not very susceptible and do not serve as a major source of inoculum. Thus, the only conditions under which we recommend that applications even be considered is during a major bloom which will yield sufficient fruit to justify the cost of application.

Fungicide applications can protect flowers from infection and benomyl can prevent symptom development if applied up to 48 h after infection begins. At the end of the bloom where there are no longer any unopened flowers present, sprays are likewise not justified. Regardless of what has occurred during the main bloom, applications at this time will no longer control the disease or increase yield.

4. Prebloom risk: The most important factors in determining whether an epidemic of PFD occurs or not are the weather during the bloom period and the amount of inoculum available at that time. Nevertheless, the geographic location of a planting and the previous history of the disease in a grove play a role in decisions. The other important factors are the inherent susceptibility of the cultivar planted and the relative frequency of flowering periods during the year. Thus, the highest risk values are assigned to the most susceptible varieties grown in tropical areas where trees tend to have two or more bloom periods per year, with lower values given to the same varieties grown in areas with single blooms. In areas with a single bloom, declining trees which flower off-season can serve to maintain inoculum levels and thus increase risk. Declining trees in areas with multiple blooms do not greatly increase disease risk. Risk values are assigned accordingly.

5. Disease history: With regard to the past occurrence of the disease in the plantings, points are assigned according to the history over the past 5 years to take into account the overall situation and to consider inoculum carried over from the previous year. Points are assigned according to the number of economically significant outbreaks over the last 5 years. Disease severity from the previous year is assessed by estimating the number of persistent calyces on the trees from the last season. The number of persistent calyces are assessed by observing 20 trees in each planting and the values assigned as shown in table. If scattered flowers that may occur prior to the main bloom become infected, inoculum can build up and increase risk for the main bloom. Risk values are added to the prebloom values as shown in table.

Table: Factors considered and points assigned to conditions occurring prior to the bloom in the development of PFD-FAD.

Factor		
	Condition	Risk value
Number of outbreaks in the last 5 years		
	None	1
	1-2	3
	3-4	7
	Every year	10
Number of buttons/tree		

	None	0
	1-100	5
	>100	10
PFD on early bloom		
	None	0
	Present	5
Varietal Susceptibility and number of blooms*		
Highly susceptible - lemons, limes, Navel, Natal, and Pera sweet oranges		
	Multiple	10
	Single	8
Moderately susceptible - Valencia orange, most tangelos		
	Multiple	10
	Single	5
Susceptible - early and mid-season oranges		
	Multiple	7
	Single	3
Tolerant - grapefruit		
	Multiple	3
	Single	1

In the case of single blooms, points are added according to the number of declining trees in the grove: None, 0; few scattered, 1; more than 10%, 2.

Varietal susceptibility and number of blooms. The citrus species and cultivar and the number of flowering periods per year are considered in conjunction with one another. In the case of a single bloom, points are added to the scores for the number of blooms and citrus variety according to the prevalence of declining trees in the planting, but no points are added for groves with multiple blooms.

The sum is taken for all of the values for:

- PFD in the last 5 years;

- Persistent calyces from the previous season;

- Cultivar susceptibility, number of blooms per tree, and the frequency of declining trees;

- The presence of PFD on the early bloom are summed. Since all prebloom risk is low compared to the risk factors occurring during the bloom, this value is divided by 7 before adding to the other risk values.

Weather Risk

1. Rainfall: Rain is needed for epidemic development to supply moisture for infection

and, as importantly, to disperse conidia by the force of droplets impacting on spore-laden petals. The amount of rain is considered in the system, but the force of the rain is as important. Fortunately, these two factors are highly related and the amount of rainfall is a good indicator of conidial dispersion.

2. Leaf wetness: At least 8 h of moisture are needed for infection and the amount of infection increases as the duration of wetness increases. We consider only the number of hours of wetness that occurred during and after a rain in the last 5 days. Without the force of rainfall, conidia are not dispersed. Dews and fogs are not considered since only localized infection results from even extended wetting periods. Dew can result in local infection of flowers from leaves or infected flowers, which increases available inoculum. This is considered under inoculum level.

Table: Weather factors considered during the bloom and points assigned in the development of PFD-FAD.

Factor				
	Condition			
		Rainfall total - last 5 days (mm)	Wetting during and after rain (h)	Risk value
Rain and leaf wetness				
		<5	<8	0
			9-16	1
			>16	2
		5-15	<8	1
			8-16	3
			>16	5
		15-25	<8	1
			8-16	4
			>16	7
		25-50	<8	1
			8-16	6
			>16	10
		>50	<8	1
			8-16	8
			>16	10
Previous rain				
	Number of rainfall events prior to last 5 days			
		0		0
		1-3		1
		>3		3

In the PFD model, very high rainfall occasionally can result in the prediction of the need for a fungicide application even when there is no inoculum present for dispersal. Similar problems occurred in the development of this advisory system. Thus, we introduced a condition such that "if disease history and the disease risk values are zero, then no application is recommended."

3. Previous Rain: Rains which occur during the bloom period but prior to the last 5 days must also be considered. Such rains bring about build-up of unseen inoculum such as large numbers of conidia on dried petals or on the surface of vegetative structures. The only rains considered are those that occur after formation of white pinhead or button bloom. Risk values are assigned according to the number of events which have occurred prior to the last 5 days.

4. Inoculum Level Risk: The inoculum availability is crucial to development of an epidemic. Propagule levels on leaves from the previous season are always low and are generally in the range of 1 to 100 propagules per leaf. In contrast, infected petals with abundant acervuli may bear as many as 1-10 million conidia per petal. Thus, if there is no PFD on the current bloom, the risk of an epidemic is small even with heavy rains and extended wetting periods. However, in order to achieve good disease control, fungicides must be applied when the number of diseased flowers are low, generally 1 to 10 infected flowers per tree. Once infection levels exceed 10 diseased flowers per tree, PFD becomes extremely difficult to control.

Table: Points assigned according to the disease situation in the grove in the development of PFD-FAD.

Condition	Risk value
No PFD evident on the current bloom	0
No PFD evident on the current bloom	0
Few, scattered foci in grove	5
Low levels (1 to 10 affected flowers per tree)	7
Many infected flowers (> 10 per tree)	10

Total risk. The total risk value for a given situation is obtained by the following formula:

Total risk score = (Prebloom risk/7) + weather risk + inoculum level risk.

5. Trigger Values: The total risk score needed to trigger a fungicide application varies according to the time of the most recent application. If fungicide has been applied recently, higher trigger values are needed.

Table: Values required to trigger a spray according to the time of the last fungicide application.

Condition	Trigger value
No fungicide last 14 days	> 11

Fungicide application 10-13 days ago	> 13
Fungicide application 7-9 days ago	> 15
Fungicide application in the last 6 days	No spray

Implementation and Output

The PFD-FAD system was implemented as a Java servlet that interacts with an HTML form. Users can specify factors and answer questions by using pull-down lists on the form. On the right side of each list, there is a button that can be used to access additional information and further explanation of each item. Once users have answered questions and specified all conditions, they submit the form. Answers are sent to the Java servlet which resides on a remote server. The servlet analyzes the user's answers and returns a recommendation in the form of an HTML page that appears on the user's Web browser. A summary of the factors submitted and a recommendation are presented.

The PFD-FAD input screen. Users make selections from pull-down lists and then submit to obtain a recommendation.

The PFD-FAD recommendation. The Java servlet generates an HTML file containing a summary of the factors entered by the user along with a recommendation.

Assessment of PFD-FAD

Field tests were conducted in 2001 at a large farm near Itapetininga, São Paulo, Brazil, to compare spray timing using PFD-FAD, the PFD model, the grower's program and an unsprayed control. Fungicides were applied to one row of trees with a guard row between treatments. Each treatment was replicated twice in each of seven blocks of trees, five of Pera sweet orange and two of Natal. The fungicides were applied with a Jacto Arbus 2000 sprayer using a spray volume of 780 liters/ha on the dates indicated in table.

Table: Effect and value of different systems for timing of fungicide applications for control of postbloom fruit drop.

Treatment	Spray-dates[a]	Buttons/12 branches /tree	Fruit/12 branches /tree	Application cost/ha	Fungicide cost/ha	Total
Control	-	91	23.9	0	0	0
PFD-FAD[b]	8/24	54	29.2	$8.70 (x1)	$32.50	$41.20
PFD Model[b]	7/27-30, 8/23-24	45	32.9	$8.70 (x2)	$65.00	$82.40
Grower program[c]	7/26-27, 8/22-24, 8/28-29	46	31.3	$8.70 (x3)	$97.50	$123.60
Least significant difference (P > 0.05)		28	5.0			

a. Ranges of spray dates indicate the timing of the sprays in the 7 different locations.

b. Fungicide used for the PFD model and PFD-FAD was Derosal (carbendazin) at 0.78 liters/ha in all cases.

c. Grower usually applied Derosal at 0.39 liters/ha + Folpan (folpet) at 0.74 kg/ha; but sometimes used Derosol at 0.59 liters/ha + Dithane PM (mancozeb) at 6.94 liter/ha, Score (difenoconazole) at 0.02 liters/ha, or Score at 0.02 liters/ha + Dithane PM at 0.78 liters/ha.

Counts of persistent calyces and fruit were made in December 2001. Ten trees were selected arbitrarily in the treated row for evaluations. Three branches about 0.7 m long were selected in each quadrant of the tree and the number of persistent calyces and fruit counted. Data were expressed as the number per 12 branches per tree. Approximate fungicide and application costs and fruit prices were obtained from local growers, distributors, and other citrus industry sources.

All spray programs significantly reduced the number of persistent calyces per tree and increased the fruit production compared to the unsprayed control. There were no significant differences among the different systems used to time applications. The PFD model reduced spray cost by more than 50% and PFD-FAD reduced costs to about one-third of that in the grower program.

Macrophomina Phaseolina

Macrophomina phaseolina is a Botryosphaeriaceae plant pathogen fungus that causes damping off, seedling blight, collar rot, stem rot, charcoal rot, basal stem rot, and root rot on many plant species.

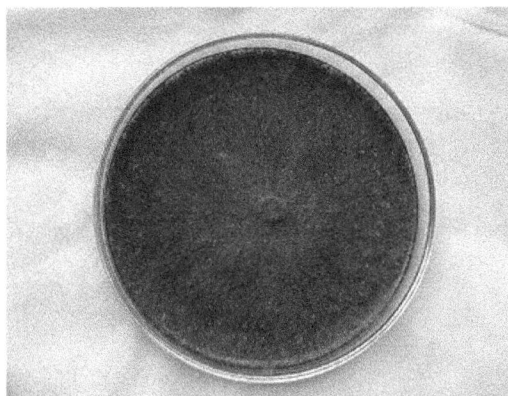

Culture of Macrophomina phaseolina on a PDA plate.

Hosts, Symptoms and Signs

One of the most harmful seed and soil borne pathogens, *Macrophomina phaseolina* is a fungus that infects nearly 500 plant species in more than 100 families. The hosts include: peanut, cabbage, pepper, chickpea, soybean, sunflower, sweet potato, alfalfa, sesame, potato, sorghum, wheat, and corn, among others. The identification of isolates of *M. phaseolina* is usually based on morphology and efforts to divide the pathogen into subspecies, but because there are wide intraspecific variations in the phenotype of the isolates, these criteria are often not reliable. The failure to correctly detect and identify *M. phaseolina* using conventional culture-based morphological techniques has led scientists to develop nucleic acid-based molecular approaches, such as highly sensitive and specific polymerase chain reaction-based methods. Researchers have also recently created species-specific oligonucleotide primers and digoxigenin-labeled probes in hopes of better identifying and detecting *M. phaseolina*.

The pathogen *M. phaseolina* affects the fibrovascular system of the roots and basal internodes of its host, impeding the transport of water and nutrients to the upper parts of the plant. As a result, progressive wilting, premature dying, loss of vigor, and reduced yield are characteristic symptoms of *M. phaseolina* infection. The fungus also causes many diseases like damping off, seedling blight, collar rot, stem rot, charcoal rot, basal stem rot, and root rot. Although brown lesions may form on the hypocotyls or emerging seedlings, many symptoms occur during or after flowering, including grey discoloration of the stem and taproots, shredding of plant tissue in the stem and top of the taproot, and hollowing of the stem. Small black dots may form beneath the epidermis of the lower stem and in the taproot, giving the stems and roots a charcoal-sprinkled appearance. When the epidermis is removed, small and black microsclerotia (a sign of the disease) may be so numerous that they give a greyish-black tint to the plant tissue. In addition, reddish-brown discoloration and black streaks can form in the pith and vascular tissues of the root and stem.

Pycnidia and sclerotia of Macrophomina phaseolina on mung bean.

Disease Cycle

M. phaseolina has a monocyclic disease cycle.

Survival

The M. phaseolina fungus has aggregates of hyphal cells, which form microsclerotia within the taproots and stems of the host plants. The microsclerotia overwinter in the soil and crop residue and are the primary source of inoculum in the spring. They have been shown to survive in the soil for up to three years. They are black, spherical or oblong structures that allow the persistence of the fungus under poor conditions, such as low soil nutrient levels and temperatures above 30 C. However, in wet soils, microsclerotia survival is significantly lower, often surviving no more than 7 to 8 weeks, and mycelium cannot survive more than 7 days. Additionally, infected seeds can carry the fungus in their seed coats. These infected seeds either do not germinate or produce seedlings that die soon after emergence.

Infection

M. phaseolina is a heat- and drought-favoring disease, producing large quantities of microsclerotia under relatively low water potentials and relatively high temperatures.

In soybeans especially, charcoal rot typically occurs when the plants are experiencing significant drought stress.

When conditions are favorable, hyphae germinate from these microsclerotia. Germination of the microsclerotia occurs throughout the growing season when temperatures are between 28 and 35 C. Microsclerotia germinate on the roots' surface, and germ tubes on the end of the microsclerotia form appresoria that penetrate the hosts' epidermal cell walls using turgor pressure or through natural openings.

The hyphae infect the roots of the host plant. Initially, the hyphae enter the cortical tissue and grow intercellularly, then infect the roots and the vascular tissue. Within the vascular tissue, mycelia and sclerotia are produced and plug the vessels. This causes the greyish-black color often observed in plants infected by *M. phaseolina*, and it also prevents water and nutrients from being transported from the roots to the upper parts of the plant. Thus, due to this systemic infection, diseased plants often wilt and die prematurely.

Management

Understanding the monocyclic disease cycle of *M. phaseolina* can help plant pathologists better understand the pathogen itself, it can help horticulturalists develop disease-resistant crops, and it can help farmers understand at what point during the growing cycle to apply fungicides or implement other management techniques.

There are several techniques currently used to manage *M. phaseolina* fungal infections. Often, fungicides are used to inhibit mycelial growth. These include thiram, iprodione, carbendazim, pyraclostrobin, fluquinconazol, tolyfluanid, and metalaxyl and penflufen + trifloxystrobin. The active ingredients carbendazim and penflufen + trifloxystrobin were shown to be the most powerful to control M. phaseolina. In this same study, the *M. phaseolina* isolate showed insensitivity to the active ingredients fluquinconazole, metalaxyl, thiram and tolyfluanid. Thus, fungicides are not necessarily an effective way to manage this fungal pathogen.

However, there are alternatives to fungicides that are especially preferred by organic farmers, such as a combination of soil solarization and organic amendment. Soil solarization is a method of using solar power for controlling pathogens in the soil by mulching the soil and covering it with a large, usually transparent polyethylene tarp to trap solar energy and heat the soil. In studies, this method has proven to be as effective as fungicides. Additionally, crop rotation can be an effective management practice. According to researchers, "Rotation out of soybeans for three years may effectively reduce microsclerotia numbers and is useful for managing charcoal rot" because "corn is not as good of a host to *M. phaseolina* as soybean so rotation with corn for three years may help reduce populations but not eliminate the pathogen from the soil." Finally, tillage practices can reduce moisture in the soil and make the environment less favorable for the pathogen.

Gibberella Fujikuroi

Gibberella fujikuroi is a fungal plant pathogen. It causes bakanae disease in rice seedlings.

Rice infected with Bakane Disease.

Another name is foolish seedling disease. It gets that name because the seeds can be infected, leading to disparate outcomes for the plant. There are not many diseases that initiate similar symptoms as bakanae. As a result, it is fairly easy to identify. However, any celebrations about identifying the disease successfully will quickly turn sour when considering the impending doom of the plants.

Hosts and Symptoms

Gibberella fujikuroi is most widely known for its disease producing capabilities in rice, but barley, millet, sugarcane and maize are also susceptible. In all infected plants, similar symptoms have been found, though rice has been most predominantly studied. The most telltale symptom of Bakane is the tall, spindly look of the plant. This is a result of the gibberellins, or growth hormones, the disease secretes. Infected plants are easy to pick out, then, as they often rise above the rest of the healthy plants with regularly secreted growth hormones. However, it is also possible that stunting may occur, along with Chlorosis of the leaves of the plant, root lesions, or empty grains of plants grown to maturity.

Disease Cycle

Gibberella fujikuroi is a polycyclic ascomycete. Thus, it is possible to, at the right stage of the lifecycle of the pathogen, see perithecia or conidia under the microscope. The pathogen overwinters in perithecia and will infect through soil in seeds that are not pre-infected. Infected seed is also a source of inoculum; conidia will germinate when planted. Infected seeds are the main way the disease is propagated. Symptoms can be observed during growing season.

Environment

Bakane is observed in all rice growing areas of the world. Thus, the pathogen thrives in the same growing conditions as rice. Rice requires a warm, wet growing season. Fields may even be flooded. The disease is known to be able to spread through water. Additionally, infected spores can be spread during harvesting. Thus, it is important to quarantine the infected seeds.

Management

There have been many management solutions put forth, with the most important and widely used being the use of treated seeds. Growers should confer with the source of their seeds and as a second measure, check the weight of the seeds they receive. Light-weight seeds are typically infected. Seeds can also be treated to prevent pathogen activation. There are two options that have been found to be successful: hot water baths and chlorine treatments. However, one cannot be sure that the pathogen will be fully neutralized when subjected to these conditions.

Resistance in rice has also been studied. Specifically, the Binam cultivar has been found to be the most resistant to the disease, thus producing the largest yield in experiments when the disease was purposefully induced. Other varieties have shown partial resistance, but none as strong as Binam.

One new option that is under investigation for management potential is treating the seeds with silver nanoparticles. The particles are a known antifungal that are not toxic to humans. In one study, treatment with the particles reduced the incidence of the disease significantly.

Importance

In rice crops, this disease has made a staggering economic impact. Losses have been specifically high in Asian countries, namely India, Thailand, and Japan. Specifically, Basmati rice has been a main target for Gibberella fujikuroi. At times, growers have lost up to 50% of their crop. However, rates of disease are not often so high and only occur during epidemic years.

Luckily, attempts at management have found success in the past and new treatments are continually researched. Growers can be confident that they will find an option to deal with this disease if it is present in their crop, if not initially, then hopefully by the next growing season.

Pathogenesis

The Gibberellins secreted by the pathogen travel systemically throughout the plant to induce the rapid growth of the hosts. Amount of Gibberellin is important in determining the extent of the disease. Another interaction between the plant and pathogen is the sporulation of mycelium at the lower levels of the plant - white fungal masses can be observed. Conidia, the secondary inoculum, are the end result of these spore masses.

Athelia Rolfsii

Athelia rolfsii is a corticioid fungus in the family Atheliaceae. It is a facultative plant pathogen and is the causal agent of "southern blight" disease in crops.

The fungus produces effused basidiocarps (fruit bodies) that are smooth and white. Microscopically, they consist of ribbon-like hyphae with clamp connections. Basidia are club-shaped, bearing four smooth, ellipsoid basidiospores, measuring 4–7 by 3–5 µm. Small, brownish sclerotia (hyphal propagules) are also formed, arising from the hyphae.

Southern Blight

Athelia rolfsii occurs in soil as a saprotroph, but can also attack living plants. It has an almost indiscriminate host range, but its capacity to form sclerotia (propagules that remain in the soil) means that it particularly attacks seasonal crops. It mostly occurs in warm soils (above 15 °C) and can be a serious pest of vegetables in tropical and sub-tropical regions (including Florida, where it was first recognized), causing "southern blight". It can also be called mustard seed fungus.

Disease Cycle

The soil-borne fungal pathogen *Athelia rolfsii* is a basidiomycete that typically exists only as mycelium and sclerotia (anamorph: *Sclerotium rolfsii*, or asexual state). It causes the disease Southern Blight and typically overwinters as sclerotia. The sclerotia is a survival structure composed of a hard rind and cortex containing hyphae and is typically considered the primary inoculum. The pathogen has a very large host range, affecting over 500 plant species (including tomato, onion, snapbean and pea) in the United States of America. The fungus attacks the host crown and stem tissues at the soil line by producing a number of compounds such as oxalic acid, in addition to enzymes that are pectinolytic and cellulytic. These compounds effectively kill plant tissue and allow the fungus to enter other areas of the plant. After gaining entry, the pathogen uses the plant tissues to produce mycelium (often forming mycelial mats), as well as additional sclerotia. Sclerotia formation occurs when conditions are especially warm and humid, primarily in the summer months in the United States of America. Susceptible plants exhibit stem lesions near the soil line, and thus often wilt and eventually die. Infection caused by Southern Blight is not considered systemic.

Environment

Athelia rolfsii typically prefers warm, humid climates (e.g. the name of the disease, Southern Blight) which is required for optimal growth (i.e. to produce mycelium and sclerotia). This makes the disease an important issue in regions such as the Southern United States of America, especially for solanaceous crops. In addition, oxygen rich and acidic soils have also been found to favor growth of the pathogen. Southern Blight can be spread (by way of sclerotia and mycelium) by contaminated farm tools and implements, irrigation systems and infected soil and plant material.

Management

Thus, management of the disease is critical, especially in agricultural regions. Although historically management has been difficult, there are several practical ways to reduce disease pressure. Simply avoiding infected fields is perhaps the most straightforward management technique given the large host range and durability of survival structures (i.e. sclerotia). However, when this is not possible, practicing proper sanitation and implementing effective crop rotations can help. Deep tillage has also been shown to reduce Southern Blight occurrence by burying infected plant tissues and creating an anaerobic environment that hinders pathogen growth. Soil solarization and certain organic amendments (e.g. composted chicken manure and rye-vetch green manure), as well as introducing certain *Trichoderma spp.* have also been shown to reduce plant death and number of sclerotia produced in the field in tomatoes. In addition to these cultural methods, chemical methods (e.g. fungicides) can also be employed. These methods all disrupt the production of mycelium and sclerotia, thus reducing the spread of disease.

Blast

Magnaporthe grisea, also known as rice blast fungus, rice rotten neck, rice seedling blight, blast of rice, oval leaf spot of graminea, pitting disease, ryegrass blast, and Johnson spot, is a plant-pathogenic fungus that causes a serious disease affecting rice. It is now known that M. grisea consists of a cryptic species complex containing at least two biological species that have clear genetic differences and do not interbreed. Complex members isolated from Digitaria have been more narrowly defined as M. grisea. The remaining members of the complex isolated from rice and a variety of other hosts have been renamed Magnaporthe oryzae. Confusion on which of these two names to use for the rice blast pathogen remains, as both are now used by different authors.

Members of the Magnaporthe grisea complex can also infect other agriculturally important cereals including wheat, rye, barley, and pearl millet causing diseases called blast disease or blight disease. Rice blast causes economically significant crop losses annually. Each year it is estimated to destroy enough rice to feed more than 60 million people. The fungus is known to occur in 85 countries worldwide.

Hosts and Symptoms

M. grisea is an ascomycete fungus. It is an extremely effective plant pathogen as it can reproduce both sexually and asexually to produce specialized infectious structures known as appressoria that infect aerial tissues and hyphae that can infect root tissues.

Lesions on rice leaves caused by infection with M. grisea.

Rice blast lesions on plant nodes.

Rice blast has been observed on rice strains M-201, M-202, M-204, M-205, M-103, M-104, S-102, L-204, Calmochi-101, with M-201 being the most vulnerable. Initial symptoms are white to gray-green lesions or spots with darker borders produced on all parts of the shoot, while older lesions are elliptical or spindle-shaped and whitish to gray with necrotic borders. Lesions may enlarge and coalesce to kill the entire leaf. Symptoms are observed on all above-ground parts of the plant. Lesions can be seen on the leaf collar, culm, culm nodes, and panicle neck node. Internodal infection of the culm occurs in a banded pattern. Nodal infection causes the culm to break at the infected node (rotten neck). It also affects reproduction by causing the host to produce fewer seeds. This is caused by the disease preventing maturation of the actual grain.

Disease Cycle

The pathogen infects as a spore that produces lesions or spots on parts of the rice plant such as the leaf, leaf collar, panicle, culm and culm nodes. Using a structure called an appressorium, the pathogen penetrates the plant. The pathogen is able to move be-tween the plant cells using its invasive hyphae to enter through plasmodesmata. *M. grisea* then sporulates from the diseased rice tissue to be dispersed as conidiospores. After overwintering in sources such as rice straw and stubble, the cycle repeats.

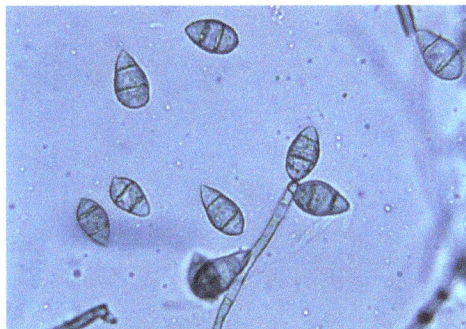

Spores of M. grisea.

A single cycle can be completed in about a week under favorable conditions where one lesion can generate up to thousands of spores in a single night. Disease lesions, however, can appear in three to four days after infection. With the ability to continue to produce the spores for over 20 days, rice blast lesions can be devastating to susceptible rice crops.

Environment

Rice blast is a significant problem in temperate regions and can be found in areas such as irrigated lowland and upland. Conditions conducive for rice blast include long pe-riods of free moisture where leaf wetness is required for infection and high humidity is common. Sporulation increases with high relative humidity and at 77-82 degrees F, spore germination, lesion formation, and sporulation are at optimum levels.

In terms of control, excessive use of nitrogen fertilization as well as drought stress in-crease rice susceptibility to the pathogen as the plant is placed in a weakened state and its defenses are low. Extended drain periods also favor infection as they aerate the soil, converting ammonium to nitrate and thus causing stress to rice crops, as well.

Management

The fungus has been able to establish resistance to both chemical treatments and genetic resistance in some types of rice developed by plant breeders. It is thought that the fungus can achieve this by genetic change through mutation. In order to most effectively control infection by *M. grisea*, an integrated management program should be implemented to avoid overuse of a single control method and fight against genetic resistance. For example, eliminating crop residue could reduce the occurrence of overwintering and discourage in-oculation in subsequent seasons. Another strategy would be to plant resistant rice varieties that are not as susceptible to infection by *M. grisea*. Knowledge of the pathogenicity of *M. grisea* and its need for free moisture suggest other control strategies such as regulated irrigation and a combination of chemical treatments with different modes of action. Managing the amount of water supplied to the crops limits spore mobility thus dampening the opportunity for infection. Chemical controls such as Carpropamid have been shown to prevent penetration of the appressoria into rice epidermal cells, leaving the grain unaffected.

J. Sendra rice affected by Magnaporthe grisea.

Importance

Rice blast is the most important disease concerning the rice crop in the world. Since rice is an important food source for much of the world, its effects have a broad range. It has been found in over 85 countries across the world and reached the United States in 1996. Every year the amount of crops lost to rice blast could feed 60 million people. Although there are some resistant strains of rice, the disease persists wherever rice is grown. The disease has never been eradicated from a region.

False Smut

False smut is a fungal disease and is more severe in the years of high rainfall:

- Individual rice grain transformed into a mass of yellow fruiting bodies,

- Growth of velvety spores that enclose floral parts,

- Infected grain has greenish smut balls with a velvety appearance,

- The smut ball appears small at first and grows gradually up to the size of 1 cm,

- It is seen in between the hulls and encloses the floral parts,

- Only few grains in a panicle are usually infected and the rest are normal,

- As the fungi growth intensifies, the smut ball bursts and becomes orange then later yellowish-green or greenish-black in color,

- Infection usually occurs during the reproductive and ripening stages, infecting a few grains in the panicle and leaving the rest healthy.

Conditions that favour disease development:

- Presence of rain and high humidity,

- Presence of soils with high nitrogen content,

- Presence of wind for dissemination of the spores from plant to plant,

- Presence of overwintering fungus as sclerotia and chlamydospores,

- Flowering stage of the rice crop.

| Infected Grain | Yellow Fruiting Body | Fungal Mass on Grains | Fungal Mass on Grains |

Identification of Pathogen

- Chlamydospore formed on the spore balls are born laterally on minute sterigmata on radial hyphae, and are spherical to elliptical, warty, olivaceous, 3-5 x 4-6 μm. Younger spores are smaller, paler, and almost smooth.

- Some of the green spore balls develop one to four sclerotia in the center. These sclerotia overwinter in the field and produce stalked stromata the following summer or autumn.

- In temperate regions, the fungus survives the winter by means of sclerotia as well as chlamydospores.

- It is believed that the primary infections are initiated mainly by the ascospores produced from the sclerotia.

- Chlamydospores play an important role in secondary infection, which is a major part of the disease cycle.

Management

Cultural Methods

- Among the cultural control, destruction of straw and stubble from infected plants is recommended to reduce the disease.

- Use varieties that are found to be resistant or tolerant against the disease in India.

- Avoid field activities when the plants are wet.

- Early planted crop has less smut balls than the late planted crop.

- At the time of harvesting, diseased plants should be removed and destroyed so that sclerotia do not fall in the field. This will reduce primary inoculum for the next crop.

- Field bunds and irrigation channels should be kept clean to eliminate alternate hosts.

- Excess application of nitrogenous fertilizer should be avoided.

- Regular monitoring of disease incidence during rabi season is very essential.

- Proper Destruction of straw and stubble.

Preventive Methods

- Use of disease-free seeds that are selected from healthy crop.

- Split application of nitrogen is recommended.

- Removal and proper disposal of infected plant debris.

Chemical Methods

- Seed treatment with carbendazim 2.0g/kg of seeds.

- Spraying of copper oxychloride @ 2.5 g/litre or Propiconazole @ 1.0 ml/litre at boot leaf and milky stages will be more useful to prevent the fungal infection.

- At tillering and preflowering stages, spray Hexaconazole @ 1ml/lit or Chloro-thalonil 2g/lit.

- In areas where the disease may cause yield loss, applying captan or captafol or fentin hydroxide or mancozeb can inhibit conidial germination.

- At tillering and pre-flowering stages, spraying of carbendazim fungicide or cop-per base fungicide can effectively control the disease.

Virus Diseases

Rice Tungro Virus

Rice tungro bacilliform virus (RTBV) is a plant pararetrovirus of the family Caulimoviridae:

- Plants affected by tungro exhibit stunting and reduced tillering. Leaves become yellow or orange-yellow, it may also have rust-colored spots,

- Discoloration begins from leaf tip and extends down to the blade or the lower leaf portion,

- Delayed flowering, - panicles are small and not completely exerted,

- Most panicles are sterile or partially filled grains,

- Tungro virus disease affects all growth stages of the rice plant specifically high at vegetative stage.

Special detection technique:

- Collect leaf samples at 6 a.m.,

- The top 10 cm portion of the leaf is immersed in a solution containing 2 g of iodine and 6 g of potassium iodide in 100 ml of water for 15 minutes or 10 ml of tincture of iodine + 140 ml of water for one hour. Washed in water for detection,

- Tungro infected leaves develop dark blue streaks.

Factors favouring disease development:

- Presence of the virus sources.

- Presence of the vector.

- Age and susceptibility of host plants.

- Synchronization of the three above factors.

- All growth stages of rice plant specifically the vegetative stage.

Yellow Leaf

Infected Field

Infected and Non-Infected Field

Identification of Pathogen

- Tungro virus disease is transmitted by leafhoppers, wherein the most efficient vector is the green leafhopper,Nephotettix virescens (Distant). The disease complex is associated with rice tungro baciliform virus (RTBV) and rice tungro spherical virus (RTSV). RTBV cannot be transmitted by leafhoppers unless RTSV is present.

- Insects could acquire the virus from any part of the infected plant. After acquiring the virus, the vector can immediately transmit to the plants.

- RTBV particles are rod-shaped and 100-300 nm in length and 30-35 nm in width. It contains DNA of 8.3 kb. RTSV particles are isometric and 30 nm in diameter. It has a polyadenylated single-stranded RNA of about 12 kb.

Management

Trap Methods

- Light traps are to be set up to attract and control the leaf hopper vectors as well as to monitor the population.

- In the early morning, the population of leafhopper alighting near the light trap should be killed by spraying/dusting the insecticides. This should be practiced every day.

Cultural Methods

- Planting of resistant varieties against tungro virus disease is the most economical means of managing the disease.

- Use resistant varieties like IR 36, IR 50, ADT 37, Ponmani, Co 45, Co 48, Surekha, Vikramarya, Bharani, IR 36 and white ponni.

- Among the cultural management practices, adjusting the date of planting is recommended.

- Likewise, observing a fallow period of at least a month to eliminate hosts and viruses and vectors of the disease.

- In epidemic areas follow rotation with pulses or oil seeds.

- Apply neem cake @ 12.5 kg/20 cent nursery as basal dose.

- Plouging and harrowing the field to destroy stubbles right after harvest.

- Eradicate of other tungro hosts are also advisable.

- Destruction of weed hosts on bunds.

Chemical Methods

- Leaf yellowing can be minimized by spraying 2 % urea mixed with Mancozeb at 2.5 gm/lit.

- Instead of urea foliar fertilizer like multi-K (potassium nitrate) can be sprayed at 1 per cent which impart disease resistance because of high potassium content.

- Green leaf hoppers as vectors are to be controlled effectively in time by spraying.

- Vegetation on the bunds should also be sprayed with the insecticides. Maintain 2.5 cm of water in the nursery and broadcast anyone of the following in 20 cents Carbofuran 3 G 3.5 kg (or) Phorate 10 G 1.0 kg (or) Quinalphos 5 G 2.0 kg.

- In nursery when virus infection is low, apply Carbofuran granules @ 1 kg/ha to control vector population.

- During pre-tillering to mid-tillering when one affected hill/m is observed start application of Carbofuran granules @ 3.5kg/ha to control insect vector.

- Spray two rounds of any one of the following insecticides.

- Thiamethoxam 25 WDG 100g/ha.

- Imidacloprid 17.8 SL 100ml/ha at 15 and 30 days after transplanting. The vegetation on the bunds should also be sprayed with the insecticides.

- Special detection technique.

- PCR detection facilities available at the Department of Plant Pathology, TNAU, Coimbatore-3 can be used.

Barley Yellow Dwarf

Barley yellow dwarf is a plant disease caused by the barley yellow dwarf virus, and is the most widely distributed viral disease of cereals. It affects the economically important crop species barley, oats, wheat, maize, triticale and rice.

Biology

Barley yellow dwarf virus (BYDV) is a positive sense single-stranded RNA virus; the viron is not enveloped in a lipid coating. The virus is transmitted by aphids, and the taxonomy of the virus is based on genome organisation, serotype differences and on the primary aphid vector of each isolate.

The isolates and their major vectors (in parentheses) are:

- Subgroup I,

- ◦ MAV, a less severe strain carried by aphids (grain aphid, Sitobion avenae), SGV (Schizaphis graminum), and PAV, a less severe strain carried by aphids (bird cherry-oat aphid, Rhopalosiphum padi, grain aphid, S. avenae, and others including rose-grain aphid, Metopolophium dirhodum).

- Subgroup II, called cereal yellow dwarf virus, however CYDV is now recognised as a separate species belonging to the genus *Polerovirus* of the family Luteoviridae:

- ◦ RPV, the most severe strain carried by aphids (Bird cherry-oat aphid, Rhopalosiphum padi), RMV (Rhopalosiphum maidis).

Pathology

Wheat plants dwarfed after infection with BYDV.

When aphids feed on the phloem of the leaf, the virus is transmitted to the phloem cells. Once inside the plant, the virus begins to replicate and assemble new virions. This process requires significant metabolic input from the plant, and causes the symptoms of barley yellow dwarf disease.

The symptoms of barley yellow dwarf vary with the affected crop cultivar, the age of the plant at the time of infection, the strain of the virus, and environmental conditions, and can be confused with other disease or physiological disorders. Symptoms appear approximately 14 days after infection. Affected plants show a yellowing or reddening of leaves (on oats and some wheats), stunting, an upright posture of thickened stiff leaves, reduced root growth, delayed (or no) heading, and a reduction in yield. The heads of affected plants tend to remain erect and become black and discoloured during ripening due to colonization by saprotrophic fungi. Young plants are the most susceptible.

The virus is transmitted from the phloem when the aphid feeds. When the aphid feeds, virions go to the aphid's hind gut, the coat protein of the virus is recognised by the hind-gut epithelium, and the virion is allowed to pass into the insect's hemolymph, where it can remain indefinitely, but the virus cannot reproduce inside the aphid. The virus is actively transported into the accessory salivary gland to be released into salivary canals and ducts. The virus is then excreted in the aphid saliva during its next feeding.

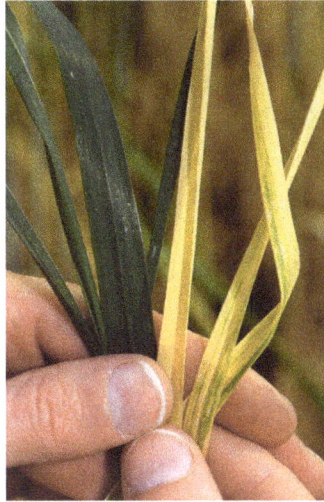

Infected wheat leaves have a reduced
ability to photosynthesise.

The host range of BYDVs consists of more than 150 species in the Poaceae; a large number of grasses both annual and perennial are alternate hosts to BYVD and can serve as reservoirs of the virus.

Sources and Spread

There are two main sources by which a cereal crop might be infected:

- By non-migrant wingless aphids already present in the field and which colonise newly-emerging crops. This is known as "green-bridge transfer".

- By winged aphids migrating into crops from elsewhere. These then reproduce and the offspring spread to neighbouring plants.

Effect on Yield

This is variable since it depends on viral strain, time of infection and rate of spread. Most severe losses are from early infections and can be as high as 50%.

Control

"Green bridge" sources must be ploughed in as early as possible. Alternatively, a desiccant herbicide should be applied 10 days prior to cultivation. Insecticide sprays may be used at crop emergence.

Drilling dates prior to mid-October favors attacks from winged migrant aphids. However, yield penalties may be experienced from late drilling. Insecticide sprays in this instance are therefore aimed at killing the aphids before significant spread can occur.

Ragged Stunt Virus

Rice ragged stunt virus (RRSV) is a plant pathogenic virus of the family Reoviridae.

Symptoms

- Infected plants severely stunted during early growth stages of the crop,
- Leaves short and dark green with serrated edges,
- Leaf blades twisted at the apex or base, which result in the spiral shape of the leaves,
- Leaf edges uneven and the twisting give the leaves a ragged appearance,
- Ragged portions of the leaves are yellow to yellow-brown,
- Vein swellings develop on the leaf blades and sheaths,
- Swellings pale yellow or white to dark brown,
- Flag leaves twisted, malformed and shortened at booting stage,
- Flowering is delayed,
- Incomplete panicle emergence,
- Nodal branches produced at upper nodes,
- Partially exserted panicles and unfilled grains.

Identification of Pathogen

- The brown planthopper transmits the disease. The early instar nymphs of the insect are more efficient in transmitting the disease than older ones. Five-day-old nymphs are the most efficient transmitters. The virus is acquired within a feeding period of 24 hours.
- Viral particles are 63-65 nm in diameter and consist of five proteins. They are mostly found in phloem and gall cells. The genome consists of ten double-stranded RNA segments. The virus is circulative and propagative in the insect vectors.

Management

- Avoid close planting and provide 30 cm rogue spacing at every 2.5 to 3.0 m to reduce the pest incidence.
- There are varieties released by IRRI, which contain genes for BPH resistance, like IR26, IR64, IR36, IR56, and IR72.

- Plouging and harrowing the field to destroy stubbles right after harvest in order to eradicate other hosts.

- Apply any one of the following to control vector BPH: Phosphamidon 40 SL 1000 ml/ha (or) Phosalone 35 EC 1500 ml/ha (or) Carbaryl 10 D 25 kg/ha (or) Acephate 75 SP 625 gm/ha (or) Chlorpyriphos 20 EC 1250 ml/ha.

Potyvirus

Potyvirus is a genus of viruses, in the family Potyviridae. Plants serve as natural hosts. There are currently 158 species in this genus including the type species Potato virus Y. The genus is named after the type virus (Potato virus Y). Potyviruses account for ~30% of the currently known plant viruses. Like Begomoviruses, members of this genus may cause significant losses in agricultural, pastoral, horticultural and ornamental crops. More than 200 species of aphids spread potyviruses and most are from the subfamily Aphidinae (genera Macrosiphum and Myzus).

Virology

The virion is non-enveloped with a flexuous and filamentous nucleocapsid, 680 to 900 nanometers (nm) long and is 11–20 nm in diameter. The nucleocapsid contains ~2000 copies of the capsid protein. The symmetry of the nucleocapsid is helical with a pitch of 3.4 nm.

The genome is a linear positive sense ssRNA ranging in size from 9000–12000 bases/nucleotides. Most potyviruses have non-segmented genomes, though a number of species are bipartite. The base composition is: 21–23.51–26% G; 23–30.15–44% A; 14.9–22.41–28% C; 15.6–24.41–30.9% U.

Genus	Structure	Symmetry	Capsid	Genomic arrangement	Genomic Segmentation
Potyvirus	Filamentous		Non-enveloped	Linear	Monopartite

In the species with a single genome, at the 5' end a protein is covalently linked (the Vg protein). It encodes a single open reading frame (ORF) expressed as a 350kDa polyprotein precursor. This is processed into seven smaller proteins: P1, helper component (HC), P3, cylindrical inclusion (CI), nuclear inclusion A (NIa), nuclear inclusion B (NIb), capsid protein (CP) and two small putative proteins known as 6K1 and 6K2. The P3 cistron also encodes a second protein—P3N-PIPO—which is generated by a +2 frameshift.

Molecular Biology

Protein P1 (~33 kiloDaltons (kDa) in molecular weight) is a serine protease.

HC (~52 KDa) is a protease that is also involved in aphid transmission. As a protease it cleaves a glycine-glycine dipeptide at its own C terminus. It also interacts with eukaryotic initiation factor 4 (eIF4). It acts as a viral RNA silencing suppressor.

The function of P3 (~41 kDa) is not known. It interacts with large subunit of the ribulose-1,5-bisphosphate carboxylase/oxygenase.

CI (~71 kDa) is an RNA helicase with ATPase activity. It is also involved in membrane attachment.

NIa (~50 kDa) is cleaved into a protease (~27 kDa) and the VPg (~22 kDa) protein.

NIb (~59 kDa) is an RNA-dependent RNA polymerase.

The functions of the 6K1 (~6 kDa) is not known. 6K2 (~6 kDa) protein, having a single trans membrane domain, is accumulating in the host cellular membranes and is thought to play a role in forming the replication vesicles of the virus.

The function of the P3N-PIPO (~25 kDa) is not known but it appears to be essential. It interacts with both the large and small subunits of the ribulose-1,5-bisphosphate carboxylase/oxygenase.

The capsid protein ranges between 30 and 35 kDa in weight.

The VPg protein interacts with eukaryotic initiation factor 4E (eIF4E). This interaction appears to be essential to viral infectivity.

Two proteinases, P1 and the helper component proteinase (HC) catalyse only auto-proteolytic reactions at their respective C termini. The remaining cleavage reactions are catalysed by either trans-proteolytic or autoproteolytic mechanisms by the small nuclear inclusion protein (NIa-Pro). This latter protein is an evolutionary homology of the picornavirus 3C proteinase.

Life Cycle

Replication may occur in the cytoplasm, nuclei, chloroplasts, Golgi apparatus, cell vacuoles or more rarely in unusual sites.

Potyviruses make proteinacous inclusions in infected plant cells. These may be crystals in either the cytoplasm or in the nucleus, as amorphous X-bodies, membranous bodies, viroplasms or pinwheels. The inclusions may or may not (depending on the species) contain virions. These inclusions can be seen in the light microscope in leaf strips of infected plant tissue stained with Orange-Green (protein stain) but not Azure A (nucleic acid stain). There are four different kinds of Potyvirus inclusions.

Replication follows the positive stranded RNA virus replication model. Positive-stranded RNA virus transcription is the method of transcription. Translation takes place by -1 ribosomal frameshifting. The virus exits the host cell by tubule-guided viral movement. Plants serve as the natural host. The virus is transmitted via a vector (insects). Transmission routes are vector and mechanical.

Genus	Host details	Tissue tropism	Entry details	Release details	Replication site	Assembly site	Transmission
Potyvirus	Plants	None	Viral movement; mechanical inoculation	Viral movement	Cytoplasm	Cytoplasm	Mechanical inoculation: aphids

Geographical Spread

Agriculture was introduced into Australia in the 18th century. This introduction also included plant pathogens. Thirty eight potyvirus species have been isolated in Australia. Eighteen potyviruses have been found only in Australia and are presumed to be endemic there. The remaining twenty appear to have been introduced with agriculture.

Other Crop Diseases

Sheath Rot

Rice sheath rot is a disease complex that can be caused by various fungal and bacterial pathogens. Major pathogens associated with rice sheath rot are fungi such as Sarocladium oryzae and Fusarium sp. belonging to the Fusarium fujikuroi complex and the bacterial pathogen Pseudomonas fuscovaginae.

Leaf Sheath Discolouration

Sheath Rotting

Infected Plant

Sheath Rotting

Symptoms

- Discoloration in the flag leaf sheath,

- Rotting occurs on the leaf sheath that encloses the young panicles,

- Irregular spots or lesions, with dark reddish brown margins and gray center, discoloration in the sheath, lesions enlarge and often coalesce and may cover the entire leaf sheath,

- Severe infection causes entire or parts of young panicles to remain within the sheath,

- Unemerged panicles rot and florets turn red-brown to dark brown,

- Whitish powdery growth inside the affected sheaths and young panicles, infected panicles and grains are sterile, shriveled, partially or unfilled and discolored,

- The disease is important during the heading towards the maturity stages of the rice crop,

- Some panicles do not emerge or emerge partially,

- Favourable conditions: High amount of nitrogen, high relative humidity, and dense crop growth favors sheath rot development. The fungus grows best at 20 to 28 °C.

Identification of Pathogen

- The fungus produces white mycelium, sparsely branched, septate, and measures 1.5-2m in diameter.

- Conidiophores arising from the mycelium are slightly thicker than the vegetative hyphae, branched once or twice, each time with 3-4 branches in a whorl.

- The ultimate branches are phialides and produce conidia that are cylindrical to slightly fusiform, often somewhat curved, hyaline, smooth, single-celled, 4-9 x 1-2.5 m.

- In some cases, the fungus infects the sheath in combination with bacterial pathogens attacking the sheath and causing grain discoloration (e.g. Pseudomonas fuscovaginae).

- The fungus invades rice through the plant's stomatas and wounds and grows intercellularly in the vascular bundles and mesophyll tissues.

- The sheath rot fungus survives as a mycelium in infected residue and on seeds.

Management

Cultural Methods

- Removal of infected stubbles after harvest.

- Optimum plant spacing can reduce the disease.

- Application of potash at tillering stage is also recommended.

- Control weeds and keep field sanitation.

Preventive Method

- Bacterization with Pseudomonas fluorescens reduced sheath rot severity by 20-42%, enhanced crop growth and increased grain yield.

- Seed treatment with P. fluorescens @ of 10g/kg of seed followed by seedling dip @ of 2.5 kg products dissolved in 100 litres and dipping the seedlings of 20 cent for 30 minutes.

- Soil application of P. fluorescens @ 2.5 kg/ha after 30 days of transplanting (This product should be mixed with 50 kg of FYM/Sand and then applied).

- Foliar spray of P. fluorescens at 0.2% concentration, commencing from 45 days after transplanting at 10 days interval for 3 times depending upon the intensity of disease.

- Seed treatment with TNAU Pf 1liquid formulation @ 10 ml/kg of seeds.

- Seedling root dipping with TNAU Pf 1liquid formulation (500 ml for one hectare seedlings).

Chemical Methods

- Apply Gypsum @ 500 kg/ha at two equal splits once basally and another at active tillering stage.

- Neem oil 3% or Ipomoea leaf powder extract (25 kg/ha) or Prosopis leaf powder extract (25 kg/ha). First spray at boot leaf stage and second 15 days later.

- Seed treatment with fungicides such as Mancozeb and Benomyl effectively eliminate seedborne inoculums.

- Hexaconazole 75% WG @ 100 mg/ lit 1st spray at the time of disease appearance and 2nd spray 15 days later.

- Application of a systemic fungicide Tridemorph and insecticide, phosphamidon in combination protected the plants from sheath rot.

- At booting stage, foliar spraying with carbendazim @ 500g/ha or edifenphos or mancozeb or Chlorothalonil or Metominostrobin @ 500 ml/ha was found to reduce sheath rot.

- Foliar spraying with Benomyl and copper oxychloride were also found to be effective.

Helminthosporium Oryzae

| Brown/ Sesame Leaf Spot | Sesame Seed Spot | Infected Field | Infected Field |

Brown Spot is called as sesame leaf spot or Helminthosporiosis, attacks seedling in nursery to milky stage in main field. The disease appears first as minute brown dots later becoming cylindrical or oval to circular (resemble sesame seed). Spots measures 0.5 to 2.0 mm in breadth - coalesce to form large patches leads to leaf dries up. Infection also occurs on panicle and neck with brown colour appearance. Infected seeds show black or brown spots on glumes are covered by olivaceous velvety growth. Seedlings die and affected nurseries can be often recognized from a distance by their brownish scorched appearance. Dark brown or black spots also appear on glumes. Infected seed causes failure of seed germination, infection in nurseries leads to seedling mortality. Reduces the grain quality and weight at harvesting stage. Yield reduction is upto 50% in severe cases. Mode of Spread and survival: Infected seeds are the most common source of primary infection. The fungus also survives on collateral hosts. Infected seeds, volunteer rice, infected rice debris and weeds are the major sources of inoculums in the field. The fungus can spread from plant to plant and in the field by airborne spores. The disease is common in nutrient-deficient soils. Favourable conditions: Temperature of 25-30°C, Relative humidity above 80 per cent, excess of nitrogen aggravates the disease incidence.

Identification of Pathogen

- The somatic structures of the fungus consist of black velvety mycelial mats which are made up of prostrate hyphae and erect sporophores.

- The hyphae are abundant, branching and anastomosing. They are dark brown or olivaceous and measure 8-15 μm or more in diameter.

- The sporophores arise as lateral branches from the hyphae.

- The conidia measure 35-170 x 11-17 μm. Typical conidia are slightly curved, widest at the middle and tapering toward the hemispherical apex, where their width approximates half the median width. Mature conidia are brownish with a moderately thin peripheral wall.

Management

Cultural Methods

- As disease is seed borne, Use disease free seeds.

- Removal of alternate & collateral hosts.

- Growing resistant varieties like ADT 44, PY 4, CORH1, CO44, Cauvery, Bhavani, TPS 4 and Dhan.

- Providing proper nutrition for optimum plant growth and prevention of water stress seem to be the most important factors in controlling brown spot.

- Soils known to be low in plant-available silicon should be amended with calcium silicate slag before planting and the field should be well irrigated to avoid water stress.

Preventive Methods

- Seed treatment with Pseudomonas fluorescens @ 10g/kg of seed followed by seedling dip (select 25 sq.m areas at one corner in the main field, stagnate water upto a depth of 2.5 cm and dissolve 2.5 kg of P. fluorescens. In this stagnated water, dip 20 cent nursery seedling for 30 minutes and transplant.

- Since the fungus is seed transmitted, a hot water seed treatment (53-54 °C) for 10-12 minutes may be effective before sowing.

- The above treatment controls primary infection at the seedling stage. Presoaking the seed in cold water for 8 hours increases effectivity of the treatment.

Chemical Method

- Seed soak/seed treatment with Captan or Thiram at 4.0g/kg of seed or treat the seed with Agrosan or Ceresan 2.5 g/kg seed to ward off appearance of seedling blight stage.

- Metominostrobin @ 500ml/ha.

- Spray Mancozeb (2.0g/lit) for 2 to 3 times at 10 - 15 day intervals. Spray preferably during early hours or afternoon at flowering and post - flowering stages.

- In addition to above seed treatment with captan or thiram or chitosan or carbendazim or mancozeb has been found to reduce seedling infection. Seed treatment with tricyclazole followed by spraying of mancozeb + tricyclazole at tillering and late booting stages gave good control of the disease.

- Application of edifenphos or chitosan or iprodione or carbendazim in the field is also advisable.

Bacterial Leaf Blight

Bacterial blight is caused by Xanthomonas oryzae pv. oryzae. It causes wilting of seedlings and yellowing and drying of leaves.

Symptoms

- Water-soaked to yellowish stripes on leaf blades or starting at leaf tips with a wavy margin.

- Leaves with undulated yellowish white or golden yellow marginal necrosis, drying of leaves back from tip and curling, leaving mid rib intact are the major symptoms.

- Appearance of bacterial ooze that looks like a milky or opaque dewdrop on young lesions early in the morning.

- Severely infected leaves tend to dry quickly.

- Loss in grain yield may be up to 60%.

Factors Favouring Disease Development

- Presence of weeds.

- Presence of rice stubbles and ratoons of infected plants.

- Presence of bacteria in the rice paddy and irrigation canals.

- Warm temperature (25-30 °C), high humidity, rain and deep water.

- Severe winds, which cause wounds and over fertilization are suitable factors for the development of the disease.

- Irrigation water and splashing or windblown rain can disseminate the bacterium from plant to plant.

- The use of trimming tools for transplanting and by handling during transplanting can also trigger new infection.

- Identification of pathogen.

- The bacteria causing disease are rods, 1.2 x 0.3-0.5 μm. They are single, occasionally in pairs but not in chains.

- They are Gram negative, non-spore-forming and devoid of capsules.

- Their colonies on nutrient agar are pale yellow, circular and smooth with an entire margin. They are convex and viscid.

Management

Preventive Method

- Seed treatment with bleaching powder (100g/l) and zinc sulfate (2%) reduce bacterial blight.

- Seed treatment - seed soaking for 8 hours in Agrimycin (0.025%) and wettable ceresan (0.05%) followed by hot water treatment for 30 min at 52-54 °C.

- Seed soaking for 8 hours in ceresan (0.1%) and treat with Streptocyclin (3g in 1 litre).

- Spray neem oil 3% or NSKE 5%.

- Spray fresh cowdung extract 20% twice (starting from initial appearance of the disease and another at fortnightly interval).

- Spray fresh cowdung extract for the control of bacterial blight. Dissolve 20 g cowdung in one litre of water; allow to settle and sieve. Use supernatant liquid. (starting from initial appearance of the disease and another at fortnightly interval).

Cultural Methods

- Grow Tolerant varieties (IR 20, IR 72, PONMANI and TKM 6).

- Secure disease free seed.

- Grow nurseries preferably in isolated upland conditions.

- Avoid clipping of seedlings during transplanting.

- Balanced fertilization and avoid excess N application.

- Skip N application at booting (if disease is moderate).

- Drain the field (except at flowering stage of the crop).

- Destruction of weeds and collateral hosts.

- Avoid flow of water from affected fields.

- Maintain proper plant spacing.

Chemical Methods

- Spray Streptomycin sulphate + Tetracycline combination 300 g + Copper oxychloride 1.25kg/ha. If necessary repeat 15 days later.

- Application of bleaching powder @ 5 kg/ha in the irrigation water is recommended at the kresek stage.

- Foliar spray with copper fungicides alternatively with Streptocyclin (250 ppm) to check secondary spread.

- Two sprays of Copper hydroxide 77 WP @ 1.25 kg/ha 30 DAP & 45 DAP.

References

- Crop-disease-manual: croppro.com.au, Retrieved 19 July, 2019

- "Gsd species synonymy: sphacelotheca reiliana (j.g. kühn) g.p. clinton". Species fungorum. Cab international. Retrieved 2014-05-18

- Martinez, c. (2000). "early infection of maize roots by sporisorium reilianum f. Sp. Zeae". Protoplasma. 213 (1–2): 83–92. Doi:10.1007/bf01280508

- Smith, c. Wayne; frederiksen, richard a. (2000-12-25). Sorghum: origin, history, technology, and production. John wiley & sons. Isbn 9780471242376

- Sorghum-downy-mildew, library, agronomy, home: pioneer.com, Retrieved 6 January, 2019

- Kolmer, james a; ordonez, maria e; groth, james v (2001). Els. John wiley & sons, ltd. Doi:10.1002/9780470015902.a0021264. Isbn 9780470015902

- Fruitdrop, review: plantmanagementnetwork.org, Retrieved 25 May, 2019

- Missouri botanical garden. "crown rot of perennials (southern blight)". Missouri botanical garden. Missouri botanical garden. Retrieved 11 december 2017

- Rice-2, rice-diseases, crop-protection: tnau.ac.in, Retrieved 1 April, 2019

- Keinath, a.p.; dubose, v.b. (2017). "management of southern blight on tomato with sdhi fungicides". Crop protection. 101 (101): 29–34. Doi:10.1016/j.cropro.2017.07.013

- Rice-8, rice-diseases, crop-protection: tnau.ac.in, Retrieved 30 March, 2019

- Wilson, r. A.; talbot, n. J. (2009). "under pressure: investigating the biology of plant infection by magnaporthe oryzae". Nature reviews microbiology. 7 (3): 185–95. Doi:10.1038/nrmicro2032. Pmid 19219052

- Rice-7, rice-diseases, crop-protection: tnau.ac.in, Retrieved 5 August, 2019

- "Virus taxonomy: 2018b release" (html). International committee on taxonomy of viruses (ictv). March 2019. Retrieved 2 may 2019

- Rice-5 rice-diseases, crop-protection: tnau.ac.in, Retrieved 2 July, 2019

4

Crop Protection

The general practice that is used to protect crop yields from harmful agents like weeds, pests and plant diseases is referred to as crop protection. It includes pesticides, bio-pesticides, herbicides, insecticides, neonicotinoid and fungicides which are used to protect crops. This chapter discusses in detail about these agents of crop protection.

Crop protection is the general method or the practice of protecting the crop yields from different agents including pests, weeds, plant diseases, and other organisms that cause damage to the agricultural crops.

Apart from crops, agricultural fields would have weeds, small animals like rats, mites, insects, pests, disease-causing pathogens and frequently raided by birds. All these factors are mainly responsible for the loss or damage to the crops. Thus to yield high crop production, farmers need to protect the crop from these pests. Hence crop protection management is important before, during and after the cultivation.

Pesticides based Crop Protection

Pesticides are chemical substances that are meant to kill pests. In general, a pesticide is a chemical or a biological agent such as a virus, bacterium, antimicrobial, or disinfectant that deters, incapacitates, kills, pests.

This use of pesticides is so common that the term pesticide is often treated as synonymous with plant protection product. It is commonly used to eliminate or control a variety of agricultural pests that can damage crops and livestock and reduce farm productivity. The most commonly applied pesticides are insecticides to kill insects, herbicides to kill weeds, rodenticides to kill rodents, and fungicides to control fungi, mold, and mildew.

Biopesticide

Biopesticides, a contraction of 'biological pesticides', include several types of pest management intervention: through predatory, parasitic, or chemical relationships. The term has been associated historically with [biological control] – and by implication – the manipulation of living organisms. Regulatory positions can be influenced by public perceptions, thus:

- In the EU, biopesticides have been defined as "a form of pesticide based on mi cro-organisms or natural products".

- The US EPA states that they "include naturally occurring substances that control pests (biochemical pesticides), microorganisms that control pests (microbial pesticides), and pesticidal substances produced by plants containing added genetic material (plant-incorporated protectants) or PIPs".

They are obtained from organisms including plants, bacteria and other microbes, fungi, nematodes, *etc.* They are often important components of integrated pest management (IPM) programmes, and have received much practical attention as substitutes to synthetic chemical plant protection products (PPPs).

Types

Biopesticides can be classified into these classes:

- Microbial pesticides which consist of bacteria, entomopathogenic fungi or viruses (and sometimes includes the metabolites that bacteria or fungi produce). Entomopathogenic nematodes are also often classed as microbial pesticides, even though they are multi-cellular.

- Bio-derived chemicals. Four groups are in commercial use: pyrethrum, rotenone, neem oil, and various essential oils are naturally occurring substances that control (or monitor in the case of pheromones) pests and microbial diseases.

- Plant-incorporated protectants (PIPs) have genetic material from other species incorporated into their genetic material (*i.e.* GM crops). Their use is controversial, especially in many European countries.

- RNAi pesticides, some of which are topical and some of which are absorbed by the crop.

Biopesticides have usually no known function in photosynthesis, growth or other basic aspects of plant physiology. Instead, they are active against biological pests. Many chemical compounds have been identified that are produced by plants to protect them from pests so they are called antifeedants. These materials are biodegradable and renewable alternatives, which can be economical for practical use. Organic farming systems embraces this approach to pest control.

RNA

RNA interference is under study for possible use as a spray-on insecticide by multiple companies, including Monsanto, Syngenta, and Bayer. Such sprays do not modify the genome of the target plant. The RNA could be modified to maintain its effectiveness as target species evolve tolerance to the original. RNA is a relatively fragile molecule that generally degrades within days or weeks of application. Monsanto estimated costs to be on the order of $5/acre.

RNAi has been used to target weeds that tolerate Monsanto's Roundup herbicide. RNAi mixed with a silicone surfactant that let the RNA molecules enter air-exchange holes in the plant's surface that disrupted the gene for tolerance, affecting it long enough to let the herbicide work. This strategy would allow the continued use of glyphosate-based herbicides, but would not per se assist a herbicide rotation strategy that relied on alternating Roundup with others.

They can be made with enough precision to kill some insect species, while not harming others. Monsanto is also developing an RNA spray to kill potato beetles One challenge is to make it linger on the plant for a week, even if it's raining. The Potato beetle has become resistant to more than 60 conventional insecticides.

Monsanto lobbied the U.S. EPA to exempt RNAi pesticide products from any specific regulations (beyond those that apply to all pesticides) and be exempted from rodent toxicity, allergenicity and residual environmental testing. In 2014 an EPA advisory group found little evidence of a risk to people from eating RNA.

However, in 2012, the Australian Safe Food Foundation alleged that the RNA trigger designed to change wheat's starch content might interfere with the gene for a human liver enzyme. Supporters countered that RNA does not appear to make it past human saliva or stomach acids. The US National Honey Bee Advisory Board told EPA that using RNAi would put natural systems at "the epitome of risk". The beekeepers cautioned that pollinators could be hurt by unintended effects and that the genomes of many insects are still unknown. Other unassessed risks include ecological (given the need for sustained presence for herbicide and other applications) and the possible for RNA drift across species boundaries.

Monsanto has invested in multiple companies for their RNA expertise, including Beeologics (for RNA that kills a parasitic mite that infests hives and for manufacturing technology) and Preceres (nanoparticle lipidoid coatings) and licensed technology from Alnylam and Tekmira. In 2012 Syngenta acquired Devgen, a European RNA partner. Startup Forrest Innovations is investigating RNAi as a solution to citrus greening disease that in 2014 caused 22 percent of oranges in Florida to fall off the trees.

Examples:

Bacillus thuringiensis, a bacterial disease of Lepidoptera, Coleoptera and Diptera, is a well-known insecticide example. The toxin from *B. thuringiensis* (Bt toxin) has been incorporated directly into plants through the use of genetic engineering. The use of Bt Toxin is particularly controversial. Its manufacturers claim it has little effect on other organisms, and is more environmentally friendly than synthetic pesticides.

Other microbial control agents include products based on:

- Entomopathogenic fungi (e.g. Beauveria bassiana, isaria fumosorosea, lecanicillium and metarhizium spp.),

- Plant disease control agents: include trichoderma spp. And ampelomyces quisqualis (a hyper-parasite of grape powdery mildew); bacillus subtilis is also used to control plant pathogens,

- Beneficial nematodes attacking insect (e.g. Steinernema feltiae) or slug (e.g. Phasmarhabditis hermaphrodita) pests,

- Entomopathogenic viruses (e.g. Cydia pomonella granulovirus),

- Weeds and rodents have also been controlled with microbial agents.

Various naturally occurring materials, including fungal and plant extracts, have been described as biopesticides. Products in this category include:

- Insect pheromones and other semiochemicals,

- Fermentation products such as Spinosad (a macro-cyclic lactone),

- Chitosan: a plant in the presence of this product will naturally induce systemic resistance (ISR) to allow the plant to defend itself against disease, pathogens and pests,

- Biopesticides may include natural plant-derived products, which include alkaloids, terpenoids, phenolics and other secondary chemicals. Certain vegetable oils such as canola oil are known to have pesticidal properties. Products based on extracts of plants such as garlic have now been registered in the EU and elsewhere.

Applications

Biopesticides are biological or biologically-derived agents, that are usually applied in a manner similar to chemical pesticides, but achieve pest management in an environmentally friendly way. With all pest management products, but especially microbial agents, effective control requires appropriate formulation and application.

Biopesticides for use against crop diseases have already established themselves on a variety of crops. For example, biopesticides already play an important role in controlling downy mildew diseases. Their benefits include: a 0-Day Pre-Harvest Interval, the ability to use under moderate to severe disease pressure, and the ability to use as a tank mix or in a rotational program with other registered fungicides. Because some market studies estimate that as much as 20% of global fungicide sales are directed at downy mildew diseases, the integration of biofungicides into grape production has substantial benefits in terms of extending the useful life of other fungicides, especially those in the reduced-risk category.

A major growth area for biopesticides is in the area of seed treatments and soil amendments. Fungicidal and biofungicidal seed treatments are used to control soil borne fungal pathogens that cause seed rots, damping-off, root rot and seedling blights. They can also be used to control internal seed–borne fungal pathogens as well as fungal pathogens that are on the surface of the seed. Many biofungicidal products also show capacities to stimulate plant host defence and other physiological processes that can make treated crops more resistant to a variety of biotic and abiotic stresses.

Disadvantages

- High specificity: which may require an exact identification of the pest/pathogen and the use of multiple products to be used; although this can also be an advantage in that the biopesticide is less likely to harm species other than the target.

- Often slow speed of action (thus making them unsuitable if a pest outbreak is an immediate threat to a crop).

- Often variable efficacy due to the influences of various biotic and abiotic factors (since some biopesticides are living organisms, which bring about pest/pathogen control by multiplying within or nearby the target pest/pathogen).

- Living organisms evolve and increase their resistance to biological, chemical, physical or any other form of control. If the target population is not exterminated or rendered incapable of reproduction, the surviving population can acquire a tolerance of whatever pressures are brought to bear, resulting in an evolutionary arms race.

- Unintended consequences: Studies have found broad spectrum biopesticides have lethal and nonlethal risks for non-target native pollinators such as Melipona quadrifasciata in Brazil.

Herbicides

Herbicides, also commonly known as weedkillers, are chemical substances used to control unwanted plants. Selective herbicides control specific weed species, while leaving the desired crop relatively unharmed, while non-selective herbicides (sometimes called total weedkillers in commercial products) can be used to clear waste ground, industrial and construction sites, railways and railway embankments as they kill all plant material with which they come into contact. Apart from selective/non-selective, other important distinctions include *persistence* (also known as *residual action*: how long the product stays in place and remains active), *means of uptake* (whether it is absorbed by above-ground foliage only, through the roots, or by other means), and *mechanism of action* (how it works). Historically, products such as common salt and other metal salts were used as herbicides, however these have gradually fallen out of favor and in some countries a number of these are banned due to their persistence in soil, and toxicity and groundwater contamination concerns. Herbicides have also been used in warfare and conflict.

Modern herbicides are often synthetic mimics of natural plant hormones which interfere with growth of the target plants. The term organic herbicide has come to mean herbicides intended for organic farming. Some plants also produce their own natural herbicides, such as the genus *Juglans* (walnuts), or the tree of heaven; such action of natural herbicides, and other related chemical interactions, is called allelopathy. Due to herbicide resistance - a major concern in agriculture - a number of products combine herbicides with different means of action. Integrated pest management may use herbicides alongside other pest control methods.

In the US in 2007, about 83% of all herbicide usage, determined by weight applied, was in agriculture. In 2007, world pesticide expenditures totaled about $39.4 billion; herbicides were about 40% of those sales and constituted the biggest portion, followed by insecticides, fungicides, and other types. Smaller quantities are used in forestry, pasture systems, and management of areas set aside as wildlife habitat.

Prior to the widespread use of chemical herbicides, cultural controls, such as altering soil pH, salinity, or fertility levels, were used to control weeds. Mechanical control (including tillage) was also (and still is) used to control weeds.

First Herbicides

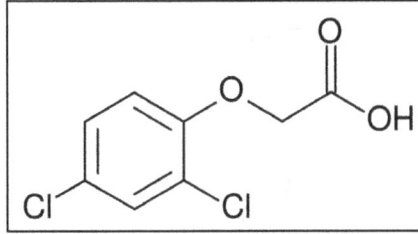

2,4-D, the first chemical herbicide, was discovered during the Second World War.

Although research into chemical herbicides began in the early 20th century, the first major breakthrough was the result of research conducted in both the UK and the US during the Second World War into the potential use of herbicides in war. The first modern herbicide, 2,4-D, was first discovered and synthesized by W. G. Templeman at Imperial Chemical Industries. In 1940, he showed that "Growth substances applied appropriately would kill certain broad-leaved weeds in cereals without harming the crops." By 1941, his team succeeded in synthesizing the chemical. In the same year, Pokorny in the US achieved this as well.

Independently, a team under Juda Hirsch Quastel, working at the Rothamsted Experimental Station made the same discovery. Quastel was tasked by the Agricultural Research Council (ARC) to discover methods for improving crop yield. By analyzing soil as a dynamic system, rather than an inert substance, he was able to apply techniques such as perfusion. Quastel was able to quantify the influence of various plant hormones, inhibitors and other chemicals on the activity of microorganisms in the soil and assess their direct impact on plant growth. While the full work of the unit remained secret, certain discoveries were developed for commercial use after the war, including the 2,4-D compound.

When 2,4-D was commercially released in 1946, it triggered a worldwide revolution in agricultural output and became the first successful selective herbicide. It allowed for greatly enhanced weed control in wheat, maize (corn), rice, and similar cereal grass crops, because it kills dicots (broadleaf plants), but not most monocots (grasses). The low cost of 2,4-D has led to continued usage today, and it remains one of the most commonly used herbicides in the world. Like other acid herbicides, current formulations use either an amine salt (often trimethylamine) or one of many esters of the parent compound. These are easier to handle than the acid.

Further Discoveries

The triazine family of herbicides, which includes atrazine, were introduced in the 1950s; they have the current distinction of being the herbicide family of greatest concern regarding groundwater contamination. Atrazine does not break down readily (within a few weeks) after being applied to soils of above neutral pH. Under alkaline soil conditions, atrazine may be carried into the soil profile as far as the water table by soil water following rainfall causing the aforementioned contamination. Atrazine is thus said to have "carryover", a generally undesirable property for herbicides.

Glyphosate (Roundup) was introduced in 1974 for nonselective weed control. Following the development of glyphosate-resistant crop plants, it is now used very extensively for selective weed control in growing crops. The pairing of the herbicide with the resistant seed contributed to the consolidation of the seed and chemistry industry in the late 1990s.

Many modern chemical herbicides used in agriculture and gardening are specifically formulated to decompose within a short period after application. This is desirable, as it allows crops and plants to be planted afterwards, which could otherwise be affected by the herbicide. However, herbicides with low residual activity (i.e., that decompose quickly) often do not provide season-long weed control and do not ensure that weed roots are killed beneath construction and paving (and cannot emerge destructively in years to come), therefore there remains a role for weedkiller with high levels of persistence in the soil.

Terminology

Herbicides are classified/grouped in various ways e.g. according to the activity, timing of application, method of application, mechanism of action, chemical family. This gives rise to a considerable level of terminology related to herbicides and their use.

Intended Outcome

- Control is the destruction of unwanted weeds, or the damage of them to the point where they are no longer competitive with the crop.

- Suppression is incomplete control still providing some economic benefit, such as reduced competition with the crop.

- Crop safety, for selective herbicides, is the relative absence of damage or stress to the crop. Most selective herbicides cause some visible stress to crop plants.

- Defoliant, similar to herbicides, but designed to remove foliage (leaves) rather than kill the plant.

Selectivity (All Plants or Specific Plants)

- Selective herbicides control or suppress certain plants without affecting the growth of other plants species. Selectivity may be due to translocation, differential absorption, physical (morphological) or physiological differences between plant species. 2,4-D, mecoprop, dicamba control many broadleaf weeds but remain ineffective against turfgrasses.

- Non-selective herbicides are not specific in acting against certain plant species and control all plant material with which they come into contact. They are used to clear industrial sites, waste ground, railways and railway embankments. Paraquat, glufosinate, glyphosate are non-selective herbicides.

Timing of Application

- Preplant: Preplant herbicides are nonselective herbicides applied to soil before planting. Some preplant herbicides may be mechanically incorporated into the soil. The objective for incorporation is to prevent dissipation through photodecomposition and/or volatility. The herbicides kill weeds as they grow through the herbicide treated zone. Volatile herbicides have to be incorporated into the soil before planting the pasture. Agricultural crops grown in soil treated with a preplant herbicide include tomatoes, corn, soybeans and strawberries. Soil fumigants like metam-sodium and dazomet are in use as preplant herbicides.

- Preemergence: Preemergence herbicides are applied before the weed seedlings emerge through the soil surface. Herbicides do not prevent weeds from germinating but they kill weeds as they grow through the herbicide treated zone by affecting the cell division in the emerging seedling. Dithopyr and pendimethalin are preemergence herbicides. Weeds that have already emerged before application or activation are not affected by pre-herbicides as their primary growing point escapes the treatment.

- Postemergence: These herbicides are applied after weed seedlings have emerged through the soil surface. They can be foliar or root absorbed, selective or non-selective, contact or systemic. Application of these herbicides is avoided during rain because the problem of being washed off to the soil makes it ineffective. 2,4-D is a selective, systemic, foliar absorbed postemergence herbicide.

Method of Application

- Soil applied: Herbicides applied to the soil are usually taken up by the root or shoot of the emerging seedlings and are used as preplant or preemergence treatment. Several factors influence the effectiveness of soil-applied herbicides. Weeds absorb herbicides by both passive and active mechanism. Herbicide adsorption to soil colloids or organic matter often reduces its amount available for weed absorption. Positioning of herbicide in correct layer of soil is very important, which can be achieved mechanically and by rainfall. Herbicides on the soil surface are subjected to several processes that reduce their availability. Volatility and photolysis are two common processes that reduce the availability of herbicides. Many soil applied herbicides are absorbed through plant shoots while they are still underground leading to their death or injury. EPTC and trifluralin are soil applied herbicides.

- Foliar applied: These are applied to portion of the plant above the ground and are absorbed by exposed tissues. These are generally postemergence herbicides and can either be translocated (systemic) throughout the plant or remain at specific site (contact). External barriers of plants like cuticle, waxes, cell wall etc. affect herbicide absorption and action. Glyphosate, 2,4-D and dicamba are foliar applied herbicide.

Persistence

Residual activity: An herbicide is described as having low residual activity if it is neutralized within a short time of application (within a few weeks or months) - typically this is due to rainfall, or by reactions in the soil. An herbicide described as having high residual activity will remain potent for a long term in the soil. For some compounds, the residual activity can leave the ground almost permanently barren.

Mechanism of Action

Herbicides are often classified according to their site of action, because as a general rule, herbicides within the same site of action class will produce similar symptoms on susceptible plants. Classification based on site of action of herbicide is comparatively better as herbicide resistance management can be handled more properly and effectively. Classification by mechanism of action (MOA) indicates the first enzyme, protein, or biochemical step affected in the plant following application.

List of Mechanisms Found in Modern Herbicides

- ACCase inhibitors: Acetyl coenzyme A carboxylase (ACCase) is part of the first step of lipid synthesis. Thus, ACCase inhibitors affect cell membrane production in the meristems of the grass plant. The ACCases of grasses are sensitive to these herbicides, whereas the ACCases of dicot plants are not.

- ALS inhibitors: the acetolactate synthase (ALS) enzyme (also known as acetohydroxyacid synthase, or AHAS) is the first step in the synthesis of the branched-chain amino acids (valine, leucine, and isoleucine). These herbicides slowly starve affected plants of these amino acids, which eventually leads to inhibition of DNA synthesis. They affect grasses and dicots alike. The ALS inhibitor family includes various sulfonylureas (SUs) (such as Flazasulfuron and Metsulfuron-methyl), imidazolinones (IMIs), triazolopyrimidines (TPs), pyrimidinyl oxybenzoates (POBs), and sulfonylamino carbonyl triazolinones (SCTs). The ALS biological pathway exists only in plants and not animals, thus making the ALS-inhibitors among the safest herbicides.

- EPSPS inhibitors: Enolpyruvylshikimate 3-phosphate synthase enzyme (EPSPS) is used in the synthesis of the amino acids tryptophan, phenylalanine and tyrosine. They affect grasses and dicots alike. Glyphosate (Roundup) is a systemic EPSPS inhibitor inactivated by soil contact.

- Synthetic auxins inaugurated the era of organic herbicides. They were discovered in the 1940s after a long study of the plant growth regulator auxin. Synthetic auxins mimic this plant hormone. They have several points of action on the cell membrane, and are effective in the control of dicot plants. 2,4-D is a synthetic auxin herbicide.

- Photosystem II inhibitors reduce electron flow from water to $NADP^+$ at the photochemical step in photosynthesis. They bind to the Qb site on the D1 protein, and prevent quinone from binding to this site. Therefore, this group of compounds causes electrons to accumulate on chlorophyll molecules. As a consequence, oxidation reactions in excess of those normally tolerated by the cell occur, and the plant dies. The triazine herbicides (including atrazine) and urea derivatives (diuron) are photosystem II inhibitors.

- Photosystem I inhibitors steal electrons from the normal pathway through FeS to Fdx to $NADP^+$ leading to direct discharge of electrons on oxygen. As a result, reactive oxygen species are produced and oxidation reactions in excess of those normally tolerated by the cell occur, leading to plant death. Bipyridinium herbicides (such as diquat and paraquat) inhibit the FeS to Fdx step of that chain, while diphenyl ether herbicides (such as nitrofen, nitrofluorfen, and acifluorfen) inhibit the Fdx to $NADP^+$ step.

- HPPD inhibitors inhibit 4-Hydroxyphenylpyruvate dioxygenase, which are involved in tyrosine breakdown. Tyrosine breakdown products are used by plants to make carotenoids, which protect chlorophyll in plants from being destroyed by sunlight. If this happens, the plants turn white due to complete loss of chlorophyll, and the plants die. Mesotrione and sulcotrione are herbicides in this class; a drug, nitisinone, was discovered in the course of developing this class of herbicides.

Herbicide Group (Labeling)

One of the most important methods for preventing, delaying, or managing resistance is to reduce the reliance on a single herbicide mode of action. To do this, farmers must know the mode of action for the herbicides they intend to use, but the relatively complex nature of plant biochemistry makes this difficult to determine. Attempts were made to simplify the understanding of herbicide mode of action by developing a classification system that grouped herbicides by mode of action. Eventually the Herbicide Resistance Action Committee (HRAC) and the Weed Science Society of America

(WSSA) developed a classification system. The WSSA and HRAC systems differ in the group designation. Groups in the WSSA and the HRAC systems are designated by numbers and letters, respectively. The goal for adding the "Group" classification and mode of action to the herbicide product label is to provide a simple and practical approach to deliver the information to users. This information will make it easier to develop educational material that is consistent and effective. It should increase user's awareness of herbicide mode of action and provide more accurate recommendations for resistance management. Another goal is to make it easier for users to keep records on which herbicide mode of actions are being used on a particular field from year to year.

Chemical Family

Detailed investigations on chemical structure of the active ingredients of the registered herbicides showed that some moieties (moiety is a part of a molecule that may include either whole functional groups or parts of functional groups as substructures; a functional group has similar chemical properties whenever it occurs in different compounds) have the same mechanisms of action. According to Forouzesh *et al.* 2015, these moieties have been assigned to the names of chemical families and active ingredients are then classified within the chemical families accordingly. Knowing about herbicide chemical family grouping could serve as a short-term strategy for managing resistance to site of action.

Use and Application

Herbicides being sprayed from the spray arms of a tractor in North Dakota.

Most herbicides are applied as water-based sprays using ground equipment. Ground equipment varies in design, but large areas can be sprayed using self-propelled sprayers equipped with long booms, of 60 to 120 feet (18 to 37 m) with spray nozzles spaced every 20–30 inches (510–760 mm) apart. Towed, handheld, and even horse-drawn sprayers are also used. On large areas, herbicides may also at times be applied aerially using helicopters or airplanes, or through irrigation systems (known as chemigation).

A further method of herbicide application developed around 2010, involves ridding the soil of its active weed seed bank rather than just killing the weed. This can successfully treat annual plants but not perennials. Researchers at the Agricultural Research Service found that the application of herbicides to fields late in the weeds' growing season greatly reduces their seed production, and therefore fewer weeds will return the following season. Because most weeds are annuals, their seeds will only survive in soil for a year or two, so this method will be able to destroy such weeds after a few years of herbicide application.

Weed-wiping may also be used, where a wick wetted with herbicide is suspended from a boom and dragged or rolled across the tops of the taller weed plants. This allows treatment of taller grassland weeds by direct contact without affecting related but desirable shorter plants in the grassland sward beneath. The method has the benefit of avoiding spray drift. In Wales, a scheme offering free weed-wiper hire was launched in 2015 in an effort to reduce the levels of MCPA in water courses.

Misuse and Misapplication

Herbicide volatilisation or spray drift may result in herbicide affecting neighboring fields or plants, particularly in windy conditions. Sometimes, the wrong field or plants may be sprayed due to error.

Use Politically, Militarily and in Conflict

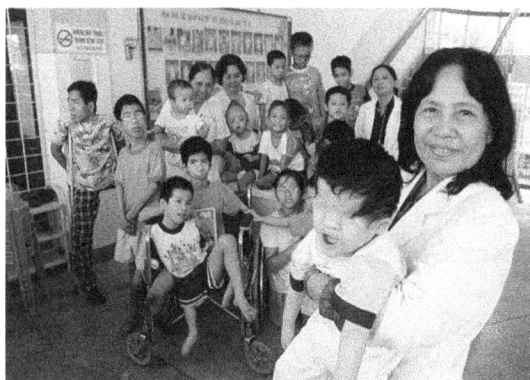

Handicapped children in Vietnam, most of them victims of Agent Orange.

Although herbicidal warfare use chemical substances, its main purpose is to disrupt agricultural food production and to destroy plants which provide cover or concealment to the enemy.

The use of herbicides as a chemical weapon by the U.S. military during the Vietnam War has left tangible, long-term impacts upon the Vietnamese people that live in Vietnam. For instance, it led to 3 million Vietnamese people suffering health problems, one million birth defects caused directly by exposure to Agent Orange, and 24% of the area of Vietnam being defoliated.

Health and Environmental Effects

Herbicides have widely variable toxicity in addition to acute toxicity arising from ingestion of a significant quantity rapidly, and chronic toxicity arising from environmental and occupational exposure over long periods. Much public suspicion of herbicides revolves around a confusion between valid statements of *acute* toxicity as opposed to equally valid statements of lack of *chronic* toxicity at the recommended levels of usage. For instance, while glyphosate formulations with tallowamine *adjuvants* are acutely toxic, their use was found to be uncorrelated with any health issues like cancer in a massive US Department of Health study on 90,000 members of farmer families for over a period of 23 years. That is, the study shows lack of chronic toxicity, but cannot question the herbicide's acute toxicity.

Some herbicides cause a range of health effects ranging from skin rashes to death. The pathway of attack can arise from intentional or unintentional direct consumption, improper application resulting in the herbicide coming into direct contact with people or wildlife, inhalation of aerial sprays, or food consumption prior to the labelled preharvest interval. Under some conditions, certain herbicides can be transported via leaching or surface runoff to contaminate groundwater or distant surface water sources. Generally, the conditions that promote herbicide transport include intense storm events (particularly shortly after application) and soils with limited capacity to adsorb or retain the herbicides. Herbicide properties that increase likelihood of transport include persistence (resistance to degradation) and high water solubility.

Phenoxy herbicides are often contaminated with dioxins such as TCDD; research has suggested such contamination results in a small rise in cancer risk after occupational exposure to these herbicides. Triazine exposure has been implicated in a likely relationship to increased risk of breast cancer, although a causal relationship remains unclear.

Herbicide manufacturers have at times made false or misleading claims about the safety of their products. Chemical manufacturer Monsanto Company agreed to change its advertising after pressure from New York attorney general Dennis Vacco; Vacco complained about misleading claims that its spray-on glyphosate-based herbicides, including Roundup, were safer than table salt and "practically non-toxic" to mammals, birds, and fish (though proof that this was ever said is hard to find). Roundup is toxic and has resulted in death after being ingested in quantities ranging from 85 to 200 ml, although it has also been ingested in quantities as large as 500 ml with only mild or moderate symptoms. The manufacturer of Tordon 101 (Dow AgroSciences, owned by the Dow Chemical Company) has claimed Tordon 101 has no effects on animals and insects, in spite of evidence of strong carcinogenic activity of the active ingredient Picloram in studies on rats.

The risk of Parkinson's disease has been shown to increase with occupational exposure to herbicides and pesticides. The herbicide paraquat is suspected to be one such factor.

All commercially sold, organic and nonorganic herbicides must be extensively tested prior to approval for sale and labeling by the Environmental Protection Agency. However, because of the large number of herbicides in use, concern regarding health effects is significant. In addition to health effects caused by herbicides themselves, commercial herbicide mixtures often contain other chemicals, including inactive ingredients, which have negative impacts on human health.

Ecological Effects

Commercial herbicide use generally has negative impacts on bird populations, although the impacts are highly variable and often require field studies to predict accurately. Laboratory studies have at times overestimated negative impacts on birds due to toxicity, predicting serious problems that were not observed in the field. Most observed effects are due not to toxicity, but to habitat changes and the decreases in abundance of species on which birds rely for food or shelter. Herbicide use in silviculture, used to favor certain types of growth following clearcutting, can cause significant drops in bird populations. Even when herbicides which have low toxicity to birds are used, they decrease the abundance of many types of vegetation on which the birds rely. Herbicide use in agriculture in Britain has been linked to a decline in seed-eating bird species which rely on the weeds killed by the herbicides. Heavy use of herbicides in neotropical agricultural areas has been one of many factors implicated in limiting the usefulness of such agricultural land for wintering migratory birds.

Frog populations may be affected negatively by the use of herbicides as well. While some studies have shown that atrazine may be a teratogen, causing demasculinization in male frogs, the U.S. Environmental Protection Agency (EPA) and its independent Scientific Advisory Panel (SAP) examined all available studies on this topic and concluded that "atrazine does not adversely affect amphibian gonadal development based on a review of laboratory and field studies."

Scientific Uncertainty of Full Extent of Herbicide Effects

The health and environmental effects of many herbicides is unknown, and even the scientific community often disagrees on the risk. For example, a 1995 panel of 13 scientists reviewing studies on the carcinogenicity of 2,4-D had divided opinions on the likelihood 2,4-D causes cancer in humans. As of 1992, studies on phenoxy herbicides were too few to accurately assess the risk of many types of cancer from these herbicides, even though evidence was stronger that exposure to these herbicides is associated with increased risk of soft tissue sarcoma and non-Hodgkin lymphoma. Furthermore, there is some suggestion that herbicides can play a role in sex reversal of certain organisms that experience temperature-dependent sex determination, which could theoretically alter sex ratios.

Resistance

Weed resistance to herbicides has become a major concern in crop production world-wide. Resistance to herbicides is often attributed to lack of rotational programmes of herbicides and to continuous applications of herbicides with the same sites of action. Thus, a true understanding of the sites of action of herbicides is essential for strategic planning of herbicide-based weed control.

Plants have developed resistance to atrazine and to ALS-inhibitors, and more recently, to glyphosate herbicides. Marestail is one weed that has developed glyphosate resistance. Glyphosate-resistant weeds are present in the vast majority of soybean, cotton and corn farms in some U.S. states. Weeds that can resist multiple other herbicides are spreading. Few new herbicides are near commercialization, and none with a molecular mode of action for which there is no resistance. Because most herbicides could not kill all weeds, farmers rotated crops and herbicides to stop resistant weeds. During its initial years, glyphosate was not subject to resistance and allowed farmers to reduce the use of rotation.

A family of weeds that includes waterhemp (Amaranthus rudis) is the largest concern. A 2008-9 survey of 144 populations of waterhemp in 41 Missouri counties revealed glyphosate resistance in 69%. Weeds from some 500 sites throughout Iowa in 2011 and 2012 revealed glyphosate resistance in approximately 64% of waterhemp samples. The use of other killers to target "residual" weeds has become common, and may be sufficient to have stopped the spread of resistance. From 2005 through 2010 researchers discovered 13 different weed species that had developed resistance to glyphosate. But since then only two more have been discovered. Weeds resistant to multiple herbicides with completely different biological action modes are on the rise. In Missouri, 43% of samples were resistant to two different herbicides; 6% resisted three; and 0.5% resisted four. In Iowa 89% of waterhemp samples resist two or more herbicides, 25% resist three, and 10% resist five.

For southern cotton, herbicide costs has climbed from between $50 and $75 per hectare a few years ago to about $370 per hectare in 2013. Resistance is contributing to a massive shift away from growing cotton; over the past few years, the area planted with cotton has declined by 70% in Arkansas and by 60% in Tennessee. For soybeans in Illinois, costs have risen from about $25 to $160 per hectare.

Dow, Bayer CropScience, Syngenta and Monsanto are all developing seed varieties resistant to herbicides other than glyphosate, which will make it easier for farmers to use alternative weed killers. Even though weeds have already evolved some resistance to those herbicides, Powles says the new seed-and-herbicide combos should work well if used with proper rotation.

Biochemistry of Resistance

Resistance to herbicides can be based on one of the following biochemical mechanisms:

- Target-site resistance: This is due to a reduced (or even lost) ability of the herbicide to bind to its target protein. The effect usually relates to an enzyme with a crucial function in a metabolic pathway, or to a component of an electron-transport system. Target-site resistance may also be caused by an overexpression of the target enzyme (via gene amplification or changes in a gene promoter).

- Non-target-site resistance: This is caused by mechanisms that reduce the amount of herbicidal active compound reaching the target site. One important mechanism is an enhanced metabolic detoxification of the herbicide in the weed, which leads to insufficient amounts of the active substance reaching the target site. A reduced uptake and translocation, or sequestration of the herbicide, may also result in an insufficient herbicide transport to the target site.

- Crossresistance: In this case, a single resistance mechanism causes resistance to several herbicides. The term target-site cross-resistance is used when the herbicides bind to the same target site, whereas non-target-site cross-resistance is due to a single non-target-site mechanism (e.g., enhanced metabolic detoxification) that entails resistance across herbicides with different sites of action.

- Multiple resistance: In this situation, two or more resistance mechanisms are present within individual plants, or within a plant population.

Resistance Management

Worldwide experience has been that farmers tend to do little to prevent herbicide resistance developing, and only take action when it is a problem on their own farm or neighbor's. Careful observation is important so that any reduction in herbicide efficacy can be detected. This may indicate evolving resistance. It is vital that resistance is detected at an early stage as if it becomes an acute, whole-farm problem, options are more limited and greater expense is almost inevitable. Table lists factors which enable the risk of resistance to be assessed. An essential pre-requisite for confirmation of resistance is a good diagnostic test. Ideally this should be rapid, accurate, cheap and accessible. Many diagnostic tests have been developed, including glasshouse pot assays, petri dish assays and chlorophyll fluorescence. A key component of such tests is that the response of the suspect population to a herbicide can be compared with that of known susceptible and resistant standards under controlled conditions. Most cases of herbicide resistance are a consequence of the repeated use of herbicides, often in association with crop monoculture and reduced cultivation practices. It is necessary, therefore, to modify these practices in order to prevent or delay the onset of resistance or to control existing resistant populations. A key objective should be the reduction in selection pressure. An integrated weed management (IWM) approach is required, in which as many tactics as possible are used to combat weeds. In this way, less reliance is placed on herbicides and so selection pressure should be reduced.

Optimising herbicide input to the economic threshold level should avoid the unnecessary use of herbicides and reduce selection pressure. Herbicides should be used to their greatest potential by ensuring that the timing, dose, application method, soil and climatic conditions are optimal for good activity. In the UK, partially resistant grass weeds such as *Alopecurus myosuroides* (blackgrass) and *Avena* spp. (wild oat) can often be controlled adequately when herbicides are applied at the 2-3 leaf stage, whereas later applications at the 2-3 tiller stage can fail badly. Patch spraying, or applying herbicide to only the badly infested areas of fields, is another means of reducing total herbicide use.

Table: Agronomic factors influencing the risk of herbicide resistance development.

Factor	Low risk	High risk
Cropping system	Good rotation	Crop monoculture
Cultivation system	Annual ploughing	Continuous minimum tillage
Weed control	Cultural only	Herbicide only
Herbicide use	Many modes of action	Single modes of action
Control in previous years	Excellent	Poor
Weed infestation	Low	High
Resistance in vicinity	Unknown	Common

Approaches to Treating Resistant Weeds

Alternative Herbicides

When resistance is first suspected or confirmed, the efficacy of alternatives is likely to be the first consideration. The use of alternative herbicides which remain effective on resistant populations can be a successful strategy, at least in the short term. The effectiveness of alternative herbicides will be highly dependent on the extent of cross-resistance. If there is resistance to a single group of herbicides, then the use of herbicides from other groups may provide a simple and effective solution, at least in the short term. For example, many triazine-resistant weeds have been readily controlled by the use of alternative herbicides such as dicamba or glyphosate. If resistance extends to more than one herbicide group, then choices are more limited. It should not be assumed that resistance will automatically extend to all herbicides with the same mode of action, although it is wise to assume this until proved otherwise. In many weeds the degree of cross-resistance between the five groups of ALS inhibitors varies considerably. Much will depend on the resistance mechanisms present, and it should not be assumed that these will necessarily be the same in different populations of the same species. These differences are due, at least in part, to the existence of different mutations conferring target site resistance. Consequently, selection for different mutations may result in different patterns of cross-resistance. Enhanced metabolism can affect even closely related herbicides to differing degrees. For example, populations of *Alopecurus myosuroides* (blackgrass) with an enhanced metabolism mechanism show resistance to pendimethalin but not to trifluralin, despite both being dinitroanilines. This is due to differences in the vulnerability of these two herbicides to oxidative metabolism. Consequently, care is needed when trying to predict the efficacy of alternative herbicides.

Mixtures and Sequences

The use of two or more herbicides which have differing modes of action can reduce the selection for resistant genotypes. Ideally, each component in a mixture should:

* Be active at different target sites,

* Have a high level of efficacy,

* Be detoxified by different biochemical pathways,

* Have similar persistence in the soil (if it is a residual herbicide),

* Exert negative cross-resistance,

* Synergise the activity of the other component.

No mixture is likely to have all these attributes, but the first two listed are the most important. There is a risk that mixtures will select for resistance to both components in the longer term. One practical advantage of sequences of two herbicides compared with mixtures is that a better appraisal of the efficacy of each herbicide component is possible, provided that sufficient time elapses between each application. A disadvantage with sequences is that two separate applications have to be made and it is possible that the later application will be less effective on weeds surviving the first application. If these are resistant, then the second herbicide in the sequence may increase selection for resistant individuals by killing the susceptible plants which were damaged but not killed by the first application, but allowing the larger, less affected, resistant plants to survive. This has been cited as one reason why ALS-resistant *Stellaria media* has evolved in Scotland recently, despite the regular use of a sequence incorporating mecoprop, a herbicide with a different mode of action.

Herbicide Rotations

Rotation of herbicides from different chemical groups in successive years should reduce selection for resistance. This is a key element in most resistance prevention programmes. The value of this approach depends on the extent of cross-resistance, and whether multiple resistance occurs owing to the presence of several different resistance mechanisms. A practical problem can be the lack of awareness by farmers of the different groups of herbicides that exist. In Australia a scheme has been introduced in which identifying letters are included on the product label as a means of enabling farmers to distinguish products with different modes of action.

Farming Practices and Resistance

Herbicide resistance became a critical problem in Australian agriculture, after many Australian sheep farmers began to exclusively grow wheat in their pastures in the 1970s. Introduced varieties of ryegrass, while good for grazing sheep, compete intensely with wheat. Ryegrasses produce so many seeds that, if left unchecked, they can completely choke a field. Herbicides provided excellent control, while reducing soil disrupting because of less need to plough. Within little more than a decade, ryegrass and other weeds began to develop resistance. In response Australian farmers changed methods. By 1983, patches of ryegrass had become immune to Hoegrass, a family of herbicides that inhibit an enzyme called acetyl coenzyme A carboxylase.

Ryegrass populations were large, and had substantial genetic diversity, because farmers had planted many varieties. Ryegrass is cross-pollinated by wind, so genes shuffle frequently. To control its distribution farmers sprayed inexpensive Hoegrass, creating selection pressure. In addition, farmers sometimes diluted the herbicide in order to save money, which allowed some plants to survive application. When resistance appeared farmers turned to a group of herbicides that block acetolactate synthase. Once again, ryegrass in Australia evolved a kind of "cross-resistance" that allowed it to rapidly break down a variety of herbicides. Four classes of herbicides become ineffective within a few years. In 2013 only two herbicide classes, called Photosystem II and long-chain fatty acid inhibitors, were effective against ryegrass.

List of Common Herbicides

Synthetic Herbicides

- 2,4-D is a broadleaf herbicide in the phenoxy group used in turf and no-till field crop production. Now, it is mainly used in a blend with other herbicides to allow lower rates of herbicides to be used; it is the most widely used herbicide in the world, and third most commonly used in the United States. It is an example of synthetic auxin (plant hormone).

- Aminopyralid is a broadleaf herbicide in the pyridine group, used to control weeds on grassland, such as docks, thistles and nettles. It is notorious for its ability to persist in compost.

- Atrazine, a triazine herbicide, is used in corn and sorghum for control of broadleaf weeds and grasses. Still used because of its low cost and because it works well on a broad spectrum of weeds common in the US corn belt, atrazine is commonly used with other herbicides to reduce the overall rate of atrazine and to lower the potential for groundwater contamination; it is a photosystem II inhibitor.

- Clopyralid is a broadleaf herbicide in the pyridine group, used mainly in turf, rangeland, and for control of noxious thistles. Notorious for its ability to persist in compost, it is another example of synthetic auxin.

- Dicamba, a postemergent broadleaf herbicide with some soil activity, is used on turf and field corn. It is another example of a synthetic auxin.

- Glufosinate ammonium, a broad-spectrum contact herbicide, is used to control weeds after the crop emerges or for total vegetation control on land not used for cultivation.

- Fluazifop (Fuselade Forte), a post emergence, foliar absorbed, translocated grass-selective herbicide with little residual action. It is used on a very wide range of broad leaved crops for control of annual and perennial grasses.

- Fluroxypyr, a systemic, selective herbicide, is used for the control of broadleaved weeds in small grain cereals, maize, pastures, rangeland and turf. It is a synthetic auxin. In cereal growing, fluroxypyr's key importance is control of cleavers, Galium aparine. Other key broadleaf weeds are also controlled.

- Glyphosate, a systemic nonselective herbicide, is used in no-till burndown and for weed control in crops genetically modified to resist its effects. It is an example of an EPSPs inhibitor.

- Imazapyr a nonselective herbicide, is used for the control of a broad range of weeds, including terrestrial annual and perennial grasses and broadleaf herbs, woody species, and riparian and emergent aquatic species.

- Imazapic, a selective herbicide for both the pre- and postemergent control of some annual and perennial grasses and some broadleaf weeds, kills plants by inhibiting the production of branched chain amino acids (valine, leucine, and isoleucine), which are necessary for protein synthesis and cell growth.

- Imazamox, an imidazolinone manufactured by BASF for postemergence application that is an acetolactate synthase (ALS) inhibitor. Sold under trade names Raptor, Beyond, and Clearcast.

- Linuron is a nonselective herbicide used in the control of grasses and broadleaf weeds. It works by inhibiting photosynthesis.

- MCPA (2-methyl-4-chlorophenoxyacetic acid) is a phenoxy herbicide selective for broadleaf plants and widely used in cereals and pasture.

- Metolachlor is a pre-emergent herbicide widely used for control of annual grasses in corn and sorghum; it has displaced some of the atrazine in these uses.

- Paraquat is a nonselective contact herbicide used for no-till burndown and in aerial destruction of marijuana and coca plantings. It is more acutely toxic to people than any other herbicide in widespread commercial use.

- Pendimethalin, a pre-emergent herbicide, is widely used to control annual grasses and some broad-leaf weeds in a wide range of crops, including corn, soybeans, wheat, cotton, many tree and vine crops, and many turfgrass species.

- Picloram, a pyridine herbicide, mainly is used to control unwanted trees in pastures and edges of fields. It is another synthetic auxin.

- Sodium chlorate *(disused/banned in some countries)*, a nonselective herbicide, is considered phytotoxic to all green plant parts. It can also kill through root absorption.

- Triclopyr, a systemic, foliar herbicide in the pyridine group, is used to control broadleaf weeds while leaving grasses and conifers unaffected.

- Several sulfonylureas, including Flazasulfuron and Metsulfuron-methyl, which act as ALS inhibitors and in some cases are taken up from the soil via the roots.

Organic Herbicides

Recently, the term "organic" has come to imply products used in organic farming. Under this definition, an organic herbicide is one that can be used in a farming enterprise that has been classified as organic. Depending on the application, they may be less effective than synthetic herbicides and are generally used along with cultural and mechanical weed control practices.

Homemade organic herbicides include:

- Corn gluten meal (CGM) is a natural pre-emergence weed control used in turfgrass, which reduces germination of many broadleaf and grass weeds.

- Vinegar is effective for 5–20% solutions of acetic acid, with higher concentrations most effective, but it mainly destroys surface growth, so respraying to treat regrowth is needed. Resistant plants generally succumb when weakened by respraying.

- Steam has been applied commercially, but is now considered uneconomical and inadequate. It controls surface growth but not underground growth and so re-spraying to treat regrowth of perennials is needed.

- Flame is considered more effective than steam, but suffers from the same difficulties.

- D-limonene (citrus oil) is a natural degreasing agent that strips the waxy skin or cuticle from weeds, causing dehydration and ultimately death.

- Saltwater or salt applied in appropriate strengths to the rootzone will kill most plants.

Insecticides

Insecticides are pesticides that are formulated to kill, harm, repel or mitigate one or more species of insect. Insecticides work in different ways. Some insecticides disrupt the nervous system, whereas others may damage their exoskeletons, repel them or control them by some other means. They can also be packaged in various forms including sprays, dusts, gels, and baits. Because of these factors, each insecticide can pose a different level of risk to non-target insects, people, pets and the environment.

- Practicing Integrated Pest Management can significantly reduce the amount of insecticides needed to control many insect problems.

- Using more than one insecticide product in the same location can increase or decrease each one's effectiveness. It may also result in a greater risk to health and the environment.

- Broad-spectrum insecticides are effective against all insects, even the good ones. Other insecticides target certain insects. Using a targeted insecticide minimizes the risk to beneficial or non-target insects.

- Some insecticides work immediately to kill insects while others may need some time to take effect.

- Insect growth regulators like pyriproxyfen and methoprene do not kill insects; they make it impossible for exposed insects to molt (grow) or lay eggs properly.

- Insecticidal baits can be used instead of spraying large areas, especially for social insects like ants. This can decrease the risk of exposure, but remember to place baits where children and pets won't have access.

Neonicotinoid

Neonicotinoids are a class of neuro-active insecticides chemically similar to nicotine. In the 1980s Shell and in the 1990s Bayer started work on their development. The neonicotinoid family includes acetamiprid, clothianidin, imidacloprid, nitenpyram, nithiazine, thiacloprid and thiamethoxam. Imidacloprid is the most widely used insecticide in the world. Compared to organophosphate and carbamate insecticides, neonicotinoids cause less toxicity in birds and mammals than insects. Some breakdown products are also toxic to insects.

Neonicotinoid use has been linked in a range of studies to adverse ecological effects, including honey-bee colony collapse disorder (CCD) and loss of birds due to a reduction in insect populations; the findings used to be conflicting and thus controversial, but recent studies by the European Food Safety Authority (EFSA) have confirmed the risk to bees. In 2013, the European Union and a few non EU countries restricted the use of certain neonicotinoids; in 2018, the EU banned the three main neonicotinoids (clothianidin, imidacloprid and thiamethoxam) for all outdoor uses. Several states in the United States have also restricted usage of neonicotinoids out of concern for pollinators and bees. In addition, in May 2019, the Environmental Protection Agency banned 12 neonicotinoids as part of a legal settlement.

Market

Neonicotinoids have been registered in more than 120 countries. With a global turnover of €1.5 billion in 2008, they represented 24% of the global market for insecticides. After the introduction of the first neonicotinoids in the 1990s, this market has grown from €155 million in 1990 to €957 million in 2008. Neonicotinoids made up 80% of all seed treatment sales in 2008.

As of 2011, seven neonicotinoids from different companies are on the market.

Name	Company	Products	Turnover in million US$ (2009)
Imidacloprid	Bayer CropScience	Confidor, Admire, Gaucho, Advocate	1,091
Thiamethoxam	Syngenta	Actara, Platinum, Cruiser	627
Clothianidin	Sumitomo Chemical/Bayer CropScience	Poncho, Dantosu, Dantop, Belay	439
Acetamiprid	Nippon Soda	Mospilan, Assail, ChipcoTristar	276
Thiacloprid	Bayer CropScience	Calypso	112
Dinotefuran	Mitsui Chemicals	Starkle, Safari, Venom	79
Nitenpyram	Sumitomo Chemical	Capstar, Guardian	8

Agricultural Usage

Efficacy

Imidacloprid is effective against sucking insects, some chewing insects, soil insects and fleas on domestic animals. It is systemic with particular efficacy against sucking insects and has a long residual activity. Imidacloprid can be added to the water used to irrigate plants. Controlled release formulations of imidacloprid take 2–10 days to release 50% of imidacloprid in water. It is applied against soil pests, seed, timber and animal pests as well as foliar treatments.

As of 2013 neonicotinoids have been used in the U.S. on about 95 percent of corn and canola crops, the majority of cotton, sorghum, and sugar beets and about half of all soybeans. They have been used on the vast majority of fruit and vegetables, including apples, cherries, peaches, oranges, berries, leafy greens, tomatoes, and potatoes, to cereal grains, rice, nuts, and wine grapes. Imidacloprid is possibly the most widely used insecticide, both within the neonicotinoids and in the worldwide market.

Seed Coatings

In agriculture, usefulness of neonicotinoid seed treatments for pest prevention depends upon the timing of planting and pest arrival. For soybeans, neonicotinoid seed treatments typically are not effective against the soybean aphid, because the compounds break down 35–42 days after planting, and soybean aphids typically are not present or at damaging population levels before this time. Neonicotinoid seed treatments can protect yield in special cases such as late-planted fields or in areas with large infestations much earlier in the growing season. Overall yield gains are not expected from neonicotinoid seed treatments for soybean insect pests in the United States, and foliar insecticides are recommended instead when insects do reach damaging levels. Health Canada estimated that neonicotinoids provide benefits equivalent to over 3% of the national farm gate value of corn and 1.5% to 2.1% of the national farm gate value of soybean in 2013.

Mode of Action

Neonicotinoids, like nicotine, bind to nicotinic acetylcholine receptors (nAChRs) of a cell and trigger a response by that cell. In mammals, nicotinic acetylcholine receptors are located in cells of both the central nervous system and peripheral nervous systems. In insects these receptors are limited to the central nervous system. Nicotinic acetylcholine receptors are activated by the neurotransmitter acetylcholine. While low to moderate activation of these receptors causes nervous stimulation, high levels overstimulate and block the receptors, causing paralysis and death. Acetylcholinesterase breaks down acetylcholine to terminate signals from these receptors. However, acetylcholinesterase cannot break down neonicotinoids and their binding is irreversible.

Basis of Selectivity

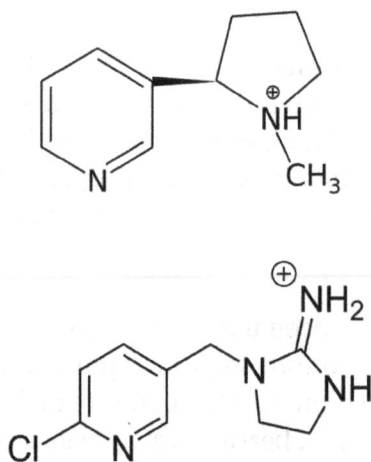

R-nicotine (top) and desnitro-imidacloprid are both protonated in the body.

Mammals and insects have different composition of the receptor subunits and the structures of the receptors. Because most neonicotinoids bind much more strongly to insect neuron receptors than to mammal neuron receptors, these insecticides are more toxic to insects than mammals.

The low mammalian toxicity of imidacloprid has been explained by its inability to cross the blood–brain barrier because of lack of a charged nitrogen atom at physiological pH. The uncharged molecule can penetrate the insect blood–brain barrier.

Other neonicotinoids have a negatively charged nitro or cyano group, which interacts with a unique, positively charged amino acid residue present on insect, but not mammalian nAChRs.

However, the breakdown product desnitro-imidacloprid, which is formed in a mammal's body during metabolism as well as in environmental breakdown, has a charged nitrogen and shows high affinity to mammalian nAChRs. Desnitro-imidacloprid is quite toxic to mice.

Toxicity

Bees

A dramatic rise in the number of annual beehive losses noticed around 2006 spurred interest in factors potentially affecting bee health. When first introduced, neonicotinoids were thought to have low toxicity to many insects, but recent research has suggested a potential toxicity to honey bees and other beneficial insects even with low levels of contact. Neonicotinoids may impact bees' ability to forage, learn and remember navigation routes to and from food sources. In lab studies, neonicotinoids were shown to increase mortality rates. Separate from lethal and sublethal effects solely due to neonicotinoid exposure, neonicotinoids are also being explored with a combination with other factors, such as mites and pathogens, as potential causes of colony collapse disorder. Neonicotinoids may be responsible for detrimental effects on bumble bee colony growth and queen production. For example, *Bombus affinis*, a bumblebee endemic to North America, has decreased in nearly 90% of its natural habitats, much of which has been attributed to the use of neonicotinoid based pesticides.

Previously undetected routes of exposure for bees include particulate matter or dust, pollen and nectar. Bees can fail to return to the hive without immediate lethality due to sub-nanogram toxicity, one primary symptom of colony collapse disorder. Separate research showed environmental persistence in agricultural irrigation channels and soil. When neonicotinoids are applied as a spray, drifting can expose bees to direct contact.

A 2012 study showed the presence of thiamethoxam and clothianidin in bees found dead in and around hives situated near agricultural fields. Other bees at the hives exhibited tremors, uncoordinated movement and convulsions, all signs of insecticide poisoning. The insecticides were also consistently found at low levels in soil up to two years after treated seed was planted and on nearby dandelion flowers and in corn pollen gathered by the bees. Insecticide-treated seeds are covered with a sticky substance to control its release into the environment, however they are then coated with talc to facilitate machine planting. This talc may be released into the environment in large amounts. Exhausted talc containing the insecticides is concentrated enough that even small amounts on flowering plants can kill foragers or be transported to the hive in contaminated pollen. Tests also showed that the corn pollen that bees were bringing back to hives tested positive for neonicotinoids at levels roughly below 100 parts per billion, an amount not acutely toxic, but enough to kill bees if sufficient amounts are consumed.

A 2012 review concluded that dietary neonicotinoids cannot be implicated in honey bee declines, but this position is provisional because important gaps remain in current knowledge.

A 2013 review concluded that neonicotinoids, as they are typically used, harm bees and that safer alternatives are urgently needed. An October 2013 study by Italian researchers demonstrated that neonicotinoids disrupt bees' immune systems, making them susceptible to viral infections to which the bees are normally resistant.

In April 2015 EASAC conducted a study of the potential effects on organisms providing a range of ecosystem services like pollination and natural pest control which are critical to sustainable agriculture. The resulting report concludes "there is an increasing body of evidence that the widespread prophylactic use of neonicotinoids has severe negative effects on non-target organisms that provide ecosystem services including pollination and natural pest control." Two studies published in *Nature* provided further evidence of the deleterious effect of neonicontinoids on bees, although the further research is needed to corroborate the findings: Oilseed rape seed coated with a combination of clothianidin and a pyrethroid "reduced wild bee density, solitary bee nesting, and bumblebee colony growth and reproduction under field conditions". In a feeding experiment, bees preferred sucrose solutions with imidacloprid or thiamethoxam, even though it "caused them to eat less food overall".

An October 2015 study demonstrated significant effects on the survival and reproductive capacities of honey bee queens exposed to neonicotinoids. Those exposed to neonicotinoids had 60% survival rates, as compared to 80% for control groups. Lower worker egg production and alterations to surviving queens' reproductive anatomy "likely corresponded to reduced queen success (alive and producing worker offspring)." The authors further claim "our study suggests that these substances [i.e., neonicotinoids] are, at least partially, responsible for harming queens and causing population declines of social bee species. Failure of queens exposed to neonicotinoids during development to successfully lay fertilised eggs that subsequently develop into workers or queens is worrisome; both castes are vital to colony survival".

A 2015 systematic review of the scientific literature on neonicotinoids and bees concluded that despite considerable research efforts, there are still significant knowledge gaps concerning the impacts of neonicotinoids on bees.

A review article concluded that while laboratory based studies have demonstrated adverse sub-lethal effects of neonicotinoid insecticides on honey bees and bumble bees, these same effects have not been observed in field studies, which is likely due to an overestimation of three key dosage factors (concentration, duration and choice) in many laboratory based studies.

A global survey covering every continent with honeybees found neonicotinoids in three-fourths of honey samples.

Other Wildlife

In March 2013, the American Bird Conservancy published a commentary on 200 studies on neonicotinoids calling for a ban on neonicotinoid use as seed treatments because of their toxicity to birds, aquatic invertebrates, and other wildlife.

A 2013 Dutch study found that water containing allowable concentrations of imidacloprid had 50% fewer invertebrate species compared with uncontaminated water. A later study found the analysis was confounded with other co-occurring insecticides and did not show imidacloprid directly affected invertebrate diversity.

In the July 2014 issue of the journal *Nature*, a study based on an observed correlation between declines in some bird populations and the use of neonicotinoid pesticides in the Netherlands demonstrated that the level of neonicotinoids detected in environmental samples correlated strongly with the decline in populations of insect-eating birds. An editorial published in the same edition found the possible link between neonicotinoid pesticide use and a decline in bird numbers "worrying", saying that the persistence of the compounds (half-life of 1000 days) and the low direct toxicity to birds themselves implies that the depletion of the birds' food source (insects) is likely responsible for the decline and that the compounds are distributed widely in the environment. The editors write that while correlation is not the same as causation, "the authors of the study also rule out confounding effects from other land-use changes or pre-existing trends in bird declines".

From June to October 2014 a comprehensive Worldwide Integrated Assessment of the impact of systemic pesticides on biodiversity and ecosystems (WIA) was published in the journal Environmental Science and Pollution Research. In a series of papers it concludes that these systemic insecticides pose a serious risk of harm to a broad range of non-target invertebrate taxa, often below the expected environmental concentrations, that their present use is not a sustainable pest management approach, and compromises the actions of numerous stakeholders in maintaining and supporting biodiversity, and this compromise subsequently negatively effects the ecological functions and services the diverse organisms perform.

Fungicides

Fungicides are biocidal chemical compounds or biological organisms used to kill parasitic fungi or their spores. A fungistatic inhibits their growth. Fungi can cause serious damage in agriculture, resulting in critical losses of yield, quality, and profit. Fungicides are used both in agriculture and to fight fungal infections in animals. Chemicals used to control oomycetes, which are not fungi, are also referred to as fungicides, as oomycetes use the same mechanisms as fungi to infect plants.

Fungicides can either be contact, translaminar or systemic. Contact fungicides are not taken up into the plant tissue and protect only the plant where the spray is deposited. Translaminar fungicides redistribute the fungicide from the upper, sprayed leaf surface to the lower, unsprayed surface. Systemic fungicides are taken up and redistributed through the xylem vessels. Few fungicides move to all parts of a plant. Some are locally systemic, and some move upwardly.

Most fungicides that can be bought retail are sold in a liquid form. A very common active ingredient is sulfur, present at 0.08% in weaker concentrates, and as high as

0.5% for more potent fungicides. Fungicides in powdered form are usually around 90% sulfur and are very toxic. Other active ingredients in fungicides include neem oil, rosemary oil, jojoba oil, the bacterium *Bacillus subtilis*, and the beneficial fungus *Ulocladium oudemansii*.

Fungicide residues have been found on food for human consumption, mostly from post-harvest treatments. Some fungicides are dangerous to human health, such as vinclozolin, which has now been removed from use. Ziram is also a fungicide that is toxic to humans with long-term exposure, and fatal if ingested. A number of fungicides are also used in human health care.

Barrier based Protection

Agrotextiles

In spite of the innumerable steps taken by Government at grass root level, the present scenario of agriculture field is characterized by shrinking of per capita water availability at a remarkable level as well as low sectoral availability of water for agricultural purposes. The problem is more pronounced due to inefficient management of water at farm level especially in areas of moderate or low rainfall. The proper utilization of monsoon water is missing to a large extent which results in depletion of water resources and low agricultural productivity. On the other hand, population is continuously increasing and there is greater stress on agricultural crops to fulfill the requirement. Under such circumstances, agriculture and horticulture fields have realized the essential need to opt for various technologies which can not only increase the agricultural productivity but also help in prevention of various natural resources like soil and water. Agro textiles are one of those techniques gaining popularity in crop and water management as well as climate control.

Agro textiles can be defined as textiles prepared by woven, knitted or non woven technique especially to be used for controlling and preserving environment and resources for various applications like agriculture, forestry, horticulture or landscape gardening. The usage of these textiles is designed to reduce the cost of production and improve the quality of produce. This will also help the farmers generate higher yield while using less resources whether natural resources like water and soil or manmade like weed killers and pesticides. Lesser usage of these chemicals will not only help in bringing the input cost down but also help to prevent environment pollution and the harmful effect of extensive usage of pesticides on soil and human health can be minimized. Agro textiles transmits light in a controlled manner and helps in maintaining the soil humidity and its temperature, prevents infestation by pests, protects crops against cold, hail storms and untimely rain thus reducing the loss chances for the farmer.

Desirable Characteristics of Agro Textiles

For a textile to be suitable for use in agriculture and horticulture it should possess certain inherent characteristics. The first and foremost is the resistance to micro organisms prevalent in soil as the textile is bound to come in contact with soil. Its susceptibility to attack by micro organisms will make it unsuitable for use as agro textile. It should also be able to withstand solar radiation and ultra violet radiation without any degradation. To prevent any side effects of wide usage of agro textiles, they should also be bio degradable. These textiles should have extensive protection properties and durability as well as a very high potential to retain water. The textiles should have good tensile strength which can be a major factor in its durability and life. These should also have good abrasion resistance properties to withstand the wear and tear of usage and climate conditions. Man made fibers are generally preferred for manufacturing of Agro Textiles because of their low cost, better durability, ease of transport and storage, although bio degradability is an issue of concern with manmade textiles. Manmade fibers like polypropylene, polyethylene, nylon and polyester are the commonly used material for making agro textiles. Various agro textiles like sun screen, wind shield, hail protection and harvesting net etc. are successfully produced using monofilament and multi filament yarns made from synthetic polymers. Utilization of non woven technique to create spun bonded fabric is extremely common for manufacturing polypropylene sheets for plant growing applications. Some natural fibers like jute and cotton are also used, where it is essential for the material to be bio degradable. However, the service life and cost of these textiles is more in comparison to synthetic substitutes. Jute is the most commonly used fiber to produce textiles for packing of agro products. Mulch mats are produced using recycled woolen fibers by using needle punched non woven technique.

These textiles are used for different aspects of crop production depending on the crop requirement and geographical location of the agriculture field. Products like sunscreen are made by warp knitted nets to protect the fields from extensive solar radiation. Knitted mono filament nets are used for production of bird protection nets to protect the crop from damage caused by birds and pests. Plant nets made of polyolefin fibers are used to prevent the decaying of fruits by keeping them away from damp soil. Polypropylene is used to produce ground cover textiles for long term weed control, moisture conservation and landscaping purpose. Wind protection nets are used to control the effect of high winds and keep out air borne sand and salt in areas close to the sea. If young plants need to be transported, root ball nets are used so that the root system is prevented from damage during transportation and replanting. To prevent insect attacks on ornamental plants and vegetables, woven and knitted polyethylene monofilament meshes are used to keep pollinating insects inside and harmful insects outside. Needle punched non woven textiles and plastic sheets are used as Mulch mats to cover the soil and help in suppressing weed growth. They can be both bio degradable and non bio degradable. Knitted Monofilament nets are used to create windbreak fences and shading screens to protect plants against the harmful effects of blustery weather and frosts. Cold and frost control fabric can be easily laid out directly on the plants for plant protection. Harvesting nets are used for certain fruits like Jamun where loss due to falling of fruit on ground is considerable. Apart from these, net fabrics are widely used for packing and transportation of farm produce.

Bird Netting

Bird netting or anti-bird netting is a form of bird pest control. It is a net used to prevent birds from reaching certain areas.

Bird protection netting comes in a variety of shapes and forms. The most common is a small mesh (1 or 2 cm squares) either extruded and bi-oriented polypropylene or woven polyethylene.

The color most used is black (as the carbon black UV inhibitor offers the best protection against solar rays), but also bird netting may be available in other colors like white (usually white netting is woven or knitted and has an even smaller mesh size as it will serve as a double purpose anti-hail net for the protection of fruits during summer hail storms or late spring during flowering) or green (usually used in home gardening and mostly sold at retail outlets for the DIY farmers).

Professional anti-bird netting comes in jumbo rolls that will offer considerable savings to the farmers or aquaculturalist. Retail chains and local stores will offer smaller packages that fit the backyard gardener's needs.

Usages

Crop Protection

Bird nets are used to prevent bird damage of vegetable and fruit crops as well as seedlings. Frugivore birds and bats can cause great damages to farmers as they tend to peck one fruit, then go to another, therefore ruining a large percentage of otherwise commercially valuable production. Once even a small portion is bitten off, that fruit cannot be sold and if harvested (even if there is no bacteria or virus brought by the frugivore) will going into rot or fermentation damaging the rest of the harvested case. Bird protection netting is applied directly on the stand-alone trees or espaliers like peaches, pears, apples, grapes, or on the side ventilation windows of growing tunnels as in the case of berries like strawberries, raspberries, blueberries, cranberries.

Dead Hedge

A dead hedge is a barrier constructed from cut branches, saplings, and foliage. The material can be from pruning, clearing, or forestry activities. Their succession is a beetle bank or hedge.

Restoration Ecology and Biological Pest Control

Dead hedges or wind-rows, as they are known in the coppice trade, are useful keeping the compartments of the coppice tidy, keeping the public from certain areas, being an excellent habitat and corridor for wildlife habitat conservation and restoration ecology, as they offer shelter for small animals, especially birds. This can be part of a beneficial "biological pest agents" habitat in biological pest control programs for natural landscapes and organic gardening.

Freshly built dead hedge after coppicing.

Agriculture

Dead hedges usually provide an enclosure for the storage of livestock. The above biological pest control dead hedges use is also part of organic farming and sustainable agriculture.

Dead hedges also recycle-reuse biomass without energy use to transport to landfills, or by burning, reducing a carbon footprint impact.

Bird Scarers

Bird scarers are a number of devices designed to scare birds, usually employed by farmers to dissuade birds from eating recently planted arable crops.

They are also used on airfields to prevent birds accumulating near runways and causing a potential hazard to aircraft.

Visual scarers

Scarecrow

One of the oldest designs of bird scarer is the scarecrow which is in the shape of a human figure. The scarecrow idea has been built upon numerous times, and not all visual scare devices are shaped like humans. The "Flashman Birdscarer," Iridescent tape, "TerrorEyes" balloons, and other visual deterrents are all built on the idea of visually scaring birds. This method doesn't work so well with all species, considering that some species frequently perch on scarecrows.

Hawk Kite

A stationary modelled owl used as a bird scarer.

Many species of bird are also naturally afraid of predators such as birds of prey. "Hawk kites" are designed to fly from poles in the wind and hover above the field to be protected. They are shaped to match the silhouette of a bird of prey.

Helikites

The Helikite bird scarer is a lighter-than-air combination of a helium balloon and a kite. Helikites fly up to 200vft in the air with or without wind. Although they do not look like hawks, they fly and hover high in the sky behaving like birds of prey. Helikites successfully exploit bird pests' instinctive fear of hawks and can reliably protect large areas of farmland.

Lasers

The use of lasers can be an effective method of bird scaring, although there is some evidence to suggest some birds are "laser-resistant". As the effectiveness of the laser decreases with increasing light levels, it is likely to be most effective at dawn and dusk. Although some lasers prove to be effective during daylight hours.

The method relies on birds being startled by the strong contrast between the ambient light and the laser beam. During low light conditions this technique is very selective and can be attuned to frequencies and wavelengths that individual bird species don't like, but at night the light beam is visible over a large distance and can cause widespread (non-species specific) disturbance. Lasers use can be limited due to safety concerns of the beam and some nations have laws which prohibit lasers above a certain power from being used. Lasers should be looked at as an additional scaring system to add to a wildlife management program.

Manually operated laser torches and automated laser bird deterrent robots that move the laser automatically towards the birds are available on the market.

Dead birds

The use of model or actual dead birds is used to signal danger to others. Initially, birds often approach the corpse but usually leave when they see the unnatural position of the bird. This approach has been frequently used in attempts to deter gulls from airports. Pheasant feed sacks often have an image of an owl with large eyes so that when empty they can be strung up to scare predators.

Balloons

An example of a visual bird-scare balloon.

Balloons are an inexpensive deterrent. However, this method relies on the movement of balloons, which is something that birds can become used to. The addition of eye illustrations on the balloons has been shown to increase this method's effectiveness as it combats the birds' ability to adapt. Commercially available "scare-eye" balloons have holographic eyes that follow birds wherever they go. The long-term effectiveness of this method can be increased by periodically moving the placement of the scare devices.

In the United Kingdom the use of balloons is subject to approval from the Civil Aviation Authority, especially around airfields.

Auditory Scarers

Audible bird scarers use noise stimuli that makes birds uncomfortable. However, once birds realize these pose no real threats, they can easily become habituated to sounds that seemed initially frightening. If just being placed in situ and left, audible bird scarers can easily become ineffective bird control solutions, however when managed on an ongoing basis or used as part of a greater bird deterrent system, sound methods can deliver quality results.

Sōzu (Shishi Odoshi)

One very old design is the Japanese sōzu, known metonymically as a shishi odoshi (although the term *shishi odoshi* properly refers to any method of scaring wild animals, including the Western scarecrow). Instead of using a visual method to distract pests, as the scarecrow does, it uses the sound of a heavy pipe repeatedly and rhythmically hitting a rock, using water as a timing device. The sōzu is also used in Japanese popular culture to denote inordinate amounts of wealth, combined with a traditional sensibility: by design, the shishi odoshi uses copious amounts of water, meaning either a very high water bill, or that it is situated on high-value land with a stream or river running through it.

Propane Cannons

A typical propane gas gun bird scarer.

Propane scare cannons are one of the most common types of bird scarer available in Europe and America. It is a propane-powered gas gun which produces a periodic explosion. The audible bang can reach very loud volumes, in excess of 150 decibels, causing a flight reaction in birds. The similarity between a scare cannon and a 12 gauge shotgun is thought to cause a startle/fear reaction, although it is also effective against birds that have not been exposed to hunting pressure.

Birds can become habituated to the sound of regular cannon detonations, especially if it does not vary in its magnitude, pitch, or time interval. However, regularly moving the cannon, utilizing on-demand firing options, including radio control, and combining cannons with other methods of deterrents can prevent habituation.

Propane scare cannons are very loud and can be disruptive to people living nearby.

One study found that restricting cannon use to only hours when birds are active and incorporating better bird damage plans drastically reduced the number of complaints from neighbours.

Electronic Repellers

Sonic bird repellers are not effective; the birds quickly acclimate to them. Usually consisting of a central unit and several speakers, the system emits digitally recorded distress calls of birds, and, in some cases, calls of predators of the target species. Some emitters randomize pitch, magnitude, time interval, sound sequence and other factors in an attempt to prevent birds from getting used to them. Many of the sounds produced are regarded as annoying to people.

Ultrasonic Scarers

Ultrasonic devices are static sound-emitting bird deterrents, which, in theory, will annoy birds to keep them away from enclosed or semi-enclosed areas. Ultrasonic scarers are not harmful to birds, however, there is debate around birds' ability to hear these frequencies at loud enough decibels. Birds are believed to have similar hearing to humans, with studies showing birds do not hear on an ultrasonic level, meaning that ultrasonic scarers often have little or no effect in deterring birds.

Cartridge Scarers

Cartridge scarers include a wide variety of noise-producing cartridges usually fired from rockets or rope bangers, or on aerodromes from modified pistols or shotguns, which produce a loud bang and emit flashes of light. They include shellcrackers, screamer shells and whistling projectiles, exploding projectiles, bird bangers and flares. Bird banger cartridges commonly use a low explosive known as flash powder.

Cartridges are projected from a shotgun with a range of 45–90 metres (148–295 ft), or pistols with a range of approximately 25 metres (82 ft), before exploding. Bird scaring cartridges can produce noise levels of up to 160 dB at varying ranges but in some countries both the cartridges and the gun require a firearms certificate.

Pyrotechnics have proved effective in dispersing birds at airports, landfill sites, agricultural crops and aquaculture facilities. At airports in the United Kingdom, shellcrackers fired from a modified pistol are the most common means of dispersing birds, as they allow the bird controller to have some directional control over birds in flight, so they can be steered away from runways.

However, as with all similar noises, there is a high risk of birds becoming used to any pyrotechnics or cartridge explosions.

Benign Acoustic Deterrence

In 2013, Dr. John Swaddle and Dr. Mark Hinders at the College of William and Mary created a new method of deterring birds using benign sounds projected by conventional and directional (parametric) speakers. The initial objectives of the technology were to displace problematic birds from airfields to reduce bird strike risks, minimize agricultural losses due to pest bird foraging, displace nuisance birds that cause extensive repair and chronic clean-up costs, and reduce bird mortality from flying into man-made structures. The sounds, referred to as a "Sonic Net," do not have to be loud and are a combination of wave forms—collectively called "colored noise"—forming non-constructive and constructive interference with how birds talk to each other.

Technically, the Sonic Net technology is not a bird scarer, but discourages birds from flying into or spending time in the target area. The impact on the birds is similar to talking in a crowded room, and since they cannot understand each other they go somewhere else. Early tests at an aviary and initial field trials at a landfill and airfield indicate that the technology is effective and that birds do not habituate to the sound. The provisional and full patents were filed in 2013 and 2014 respectively, with further research and commercialization of the technology being ongoing.

Other

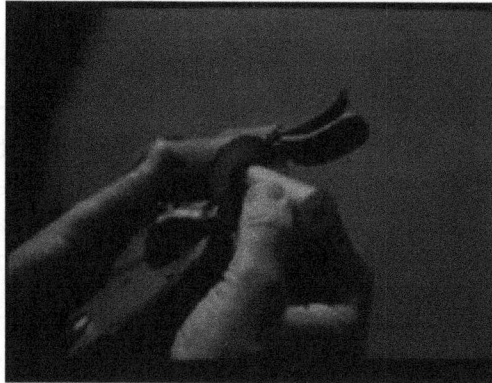

1986 U.S. Air Force video about methods to scare
birds away from airfields.

Dogs

The control of birds and other wildlife such as deer through harassment by trained border collies has been used at aerodromes, golf courses and agricultural land. The dogs represent an actual threat, and so elicit flight reactions. Habituation is unlikely as they can continually pursue and change their behaviour. Border collies are used as they are working dogs bred to herd animals and to avoid attack, and they respond well to whistle and verbal commands. A single border collie and its handler can keep an area of approximately 50 square kilometres (19.3 square miles, 4998.7 hectares, or 12,179.2

acres) free of larger birds and wildlife. However, although they are effective at deterring ground foraging birds such as waders and wildfowl, they are not so useful for species that spend most of their time flying or perching, such as raptors and swallows.

In 1999, Southwest Florida International Airport became the first commercial airport in the world to employ a border collie in an airfield wildlife control programme. After the use of the collie, numbers and species of birds on the airport declined and most birds that remained congregated in a drainage ditch away from the runway. The number of bird strikes dropped to zero compared to 13 for the same period the previous year. Several other airports and airbases have now started similar programmes.

At Dover Air Force Base, Delaware, bird strike damage to aircraft caused by birds has been reduced from an average of US$600,000 per year for the proceeding two years to US$24,000 per year after the initiation of a bird control programme that included the use of border collies.

Predators

Harris's hawk.

Using predators as a natural bird deterrent has become a recommended form of controlling bird infestations. Specially selected species are trained to deal with working in un-natural environments with distractions and dangers they would not usually encounter.

The success of this method of bird control is based on the fact that many birds have a natural fear of falcons and hawks as predators, so their presence in the area encourages problem species to disperse. The natural reaction of most prey species is to form a flock and attempt to fly above the falcon. If this fails, they will attempt to fly for cover and leave the area.

Radio Controlled Aircraft

Radio-controlled model aircraft have been used to scare or 'haze' bird pests since the early 1980s, mainly over airfields, but have also been used over agricultural areas, fisheries and landfill sites. This method has been shown to be very effective and birds habituate more slowly to a treatment in which they are being actively hazed. At Whiteman Air Force Base, Missouri, balsa wood radio-controlled aircraft are one of the primary bird harassment methods used to keep the airfield clear of raptors and other large birds, and they have also proved effective at dispersing the base's redwing blackbird roost.

Fireworks

Fireworks can also be used as bird scarers, and some jurisdictions issue special licences for agricultural fireworks. This practice has been criticised as a loophole for the sale of consumer fireworks. Again, the loud bangs can also irritate people living on nearby properties.

Combination Scarers

These combine multiple deterrents, such as using a pop up scarecrow combined with a gas gun, which in turn activates the distress call of a bird. These combination scarers are often managed by computers and synchronised across an area via the use of radio links. This synchronisation becomes more effective if there is some kind of detection system involved such as bird detecting radar.

Aerial Application

Aerial application, or what was formerly referred to as crop dusting, involves spraying crops with crop protection products from an agricultural aircraft. Planting certain types of seed are also included in aerial application. The specific spreading of fertilizer is also known as *aerial topdressing* in some countries. Many countries have severely limited aerial application of pesticides and other products because of environmental and public health hazards like spray drift; most notably, the European Union banned it outright with a few highly restricted exceptions in 2009, effectively ending the practice in all member states.

Agricultural aircraft are highly specialized, purpose-built aircraft. Today's agricultural aircraft are often powered by turbine engines of up to 1500 hp and can carry as much as 800 US gallons (3,000 l) of crop protection product. Helicopters are sometimes used, and some aircraft serve double duty as water bombers in areas prone to wildfires.

Trap Crop

A trap crop is a plant that attracts agricultural pests, usually insects, away from nearby crops. This form of companion planting can save the main crop from decimation by pests without the use of pesticides. While many trap crops have successfully divert-ed pests off of focal crops in small scale greenhouse, garden and field experiments, only a small portion of these plants have been shown to reduce pest damage at larger commercial scales. A common explanation for reported trap cropping failures, is that attractive trap plants only protect nearby plants if the insects do not move back into the main crop. In a review of 100 trap cropping examples in 2006, only 10 trap crops were classified as successful at a commercial scale, and in all successful cases, trap cropping was supplemented with management practices that specifically limited insect dispersal from the trap crop back into the main crop.

Usage

Trap crops, when used on an industrial scale, are generally planted at a key time in the pest's life-cycle, and then destroyed before that life-cycle finishes and the pest might have transferred from the trap plants to the main crop.

Examples of trap crops include:

- Alfalfa planted in strips among cotton, to draw away lygus bugs, while castor beans surround the field, or tobacco is planted in strips among it, to protect from the budworm Heliothis.

- Rose enthusiasts often plant Pelargonium geraniums among their rosebushes because Japanese beetles are drawn to the geraniums, which are toxic to them.

- Chervil is used by gardeners to protect vegetable plants from slugs.

- Rye, sesbania, and sicklepod are used to protect soybeans from corn seeding maggots, stink bugs, and velvet green caterpillars, respectively.

- Mustard and Alfalfa planted near strawberries to attract lygus bugs, a method pioneered by Jim Cochran.

Trap crops can be planted around the circumference of the field to be protected, which is assumed to act as a barrier for entry by pests, or they can be interspersed among the focul crop, for example being planted every ninth row. Planting crops in rows helps facilitate supplemental management practices that prevent insect pest dispersal back into the main field, such as driving a vehicle above the trap crop which then removes insect pests by vacuuming them off of the trap crop row or targeted insecticides, which are only deployed on the trap crop. Even if pesticides are used to control insects on the trap crop, total pesticides are greatly reduced in this scenario over conventional agri-cultural pesticide applications because they are only deployed on a small portion of the farm (the trap crop). Other strategies that prevent dispersal of insect pests back into the main crop include cutting the trap plants, applying predators to the trap plant that eat the pest, and planting a high ratio of trap plants to other plants.

Operation

Recent studies on host-plant finding have shown that flying pests are far less successful if their host-plants are surrounded by any other plant, or even "decoy-plants" made of green plastic, cardboard or any other green material. The host-plant finding process occurs in three phases.

The first phase is stimulation by odours characteristic to the host-plant. This induces the insect to try to land on the plant it seeks. But insects avoid landing on brown (bare) soil. So if only the host-plant is present, the insects will quasi-systematically find it by landing on the only green thing around. This is called an "appropriate landing". When it does an "inappropriate landing", it flies off to any other nearby patch of green. It eventually leaves the area if there are too many "inappropriate" landings.

The second phase of host-plant finding is for the insect to make short flights from leaf to leaf to assess the plant's overall suitability. The number of leaf-to-leaf flights varies according to the insect species and to the host-plant stimulus received from each leaf. But the insect must accumulate sufficient stimuli from the host-plant to lay eggs; so it must make a certain number of consecutive "appropriate" landings. Hence if it makes an "inappropriate landing", the assessment of that plant is negative and the insect must start the process anew.

Thus it was shown that clover used as a ground cover had the same disruptive effect on eight pest species from four insect orders. An experiment showed that 36% of cabbage root flies laid eggs beside cabbages growing in bare soil (which resulted in no crop), compared with only 7% beside cabbages growing in clover (which allowed a good crop). Also that simple decoys made of green card disrupted appropriate landings just as well as did the live ground cover.

References

- Crop-protection, biology: byjus.com, Retrieved 3 February, 2019

- Copping, leonard g. (2009). The manual of biocontrol agents: a world compendium. Bcpc. Isbn 978-1-901396-17-1

- Pesticides, chemistry: byjus.com, Retrieved 26 August, 2019

- Robbins, paul (2007-08-27). Encyclopedia of environment and society. Robbins, paul, 1967-, sage publications. Thousand oaks. P. 862. Isbn 9781452265582. Oclc 228071686

- van almsick, a. (2009). "new hppd-inhibitors – a proven mode of action as a new hope to solve current weed problems". Outlooks on pest management. 20: 27–30. Doi:10.1564/20feb09

- Insecticide, ptype, ingred: orst.edu, Retrieved 3 March, 2019

- Samuel fromartz (2005). Organic inc.: natural foods and how they grew. Harcourt. Isbn 978-0-15-603242-1

5
Weed Control

Weed control is a method of pest control that attempts to stop noxious weeds from competing for resources with desired flora and fauna. Various physical and mechanical methods are used to control weeds such as coverings, manual removal, thermal, tillage and seed targeting. This chapter closely examines these key concepts of weed control to provide an extensive understanding of the subject.

Weed

Weed is a general term for any plant growing where it is not wanted. Ever since humans first attempted the cultivation of plants, they have had to fight the invasion by weeds into areas chosen for crops. Some unwanted plants later were found to have virtues not originally suspected and so were removed from the category of weeds and taken under cultivation. Other cultivated plants, when transplanted to new climates, escaped cultivation and became weeds or invasive species. The category of weeds thus is ever changing, and the term is a relative one.

Weeds interfere with a variety of human activities, and many methods have been developed to suppress or eliminate them. These methods vary with the nature of the weed itself, the means at hand for disposal, and the relation of the method to the environment. Usually for financial and ecological reasons, methods used on a golf course or a public park cannot be applied on rangeland or in the forest. Herbicide chemicals sprayed on a roadside to eliminate unsightly weeds that constitute a fire or traffic hazard are not proper for use on cropland. Mulching, which is used to suppress weeds in a home garden, is not feasible on large farms. Weed control, in any event, has become a highly specialized activity. Universities and agricultural colleges teach courses in weed control, and industry provides the necessary technology. In agriculture, weed control is essential for maintaining high levels of crop production.

The many reasons for controlling weeds become more complex with the increasing development of technology. Plants become weeds as a function of time and place. Tall weeds on roadsides presumably were not problematic prior to the invention of the automobile. However, with cars and increasing numbers of drivers on roads, tall weeds became dangerous, potentially obscuring drivers' visibility, particularly at intersections. Sharp-edged grasses are nominal nuisances in a cow pasture; when the area is converted to a golf course or a public park, they become an actual nuisance. Poison oak (Toxicodendron diversilobum) is rather a pleasant shrub on a sunny hillside in the open country; in a camp ground it is a definite health hazard. Such examples could be given ad infinitum to cover every aspect of agriculture, forestry, highway, waterway and public land management, arboretum, park and golf-course care, and home landscape maintenance.

Weeds compete with crop plants for water, light, and nutrients. Weeds of rangelands and pastures may be unpalatable to animals, or even poisonous; they may cause injuries, as with lodging of foxtails (Alopecurus species) in horses' mouths; they may lower values of animal products, as in the cases of cockleburs (Xanthium species) in wool; they may add to the burden of animal care, as when horses graze in sticky tarweeds (Madia species). Many weeds are hosts of plant disease organisms. Examples are prickly lettuce (Lactuca scariola) and sow thistle (Sonchus species) that serve as hosts for downy mildew; wild mustards (Brassica species) that host clubroot of cabbage; and saltbrush (Atriplex species) and Russian thistle, in which curly top virus overwinters, to be carried to sugar beets by leafhoppers. Many weeds are hosts of insect pests, and a number are invasive species.

Ecological Significance

A dandelion is a common plant all over the world, especiallyin Europe, Asia, and the Americas. It is a well-known example of a plant that is considered a weed in some contexts (such as lawns) but not a weed in others (such as when it is used as a vegetable or herbal medicine).

Certain classes of weeds share adaptations to ruderal environments. That is to say: disturbed environments where soil or natural vegetative cover has been damaged or frequently gets damaged, disturbances that give the weeds advantages over desirable crops, pastures, or ornamental plants. The nature of the habitat and its disturbances will affect or even determine which types of weed communities become dominant.

Examples of such ruderal or pioneer species include plants that are adapted to naturally occurring disturbed environments such as dunes and other windswept areas with shifting soils, alluvial flood plains, river banks and deltas, and areas that are burned repeatedly. Since human agricultural practices often mimic these natural environments where weedy species have evolved, some weeds are effectively preadapted to grow and proliferate in human-disturbed areas such as agricultural fields, lawns, roadsides, and construction sites. The weedy nature of these species often gives them an advantage over more desirable crop species because they often grow quickly and reproduce quickly, they commonly have seeds that persist in the soil seed bank for many years, or they may have short lifespans with multiple generations in the same growing season. In contrast, perennial weeds often have underground stems that spread under the soil surface or, like ground ivy (*Glechoma hederacea*), have creeping stems that root and spread out over the ground.

Some plants become dominant when introduced into new environments because the animals in their original environment, that compete with them or feed on them are absent; in what is sometimes called the "natural enemies hypothesis", plants freed from these specialist consumers may become dominant. An example is Klamath weed, that threatened millions of hectares of prime grain and grazing land in North America after it was accidentally introduced, but was reduced to a rare roadside weed within several years after some of its natural enemies were imported during World War II. In locations where predation and mutually competitive relationships are absent, weeds have increased resources available for growth and reproduction. The weediness of some species that are introduced into new environments may be caused by their production of allelopathic chemicals which indigenous plants are not yet adapted to, a scenario sometimes called the "novel weapons hypothesis". These chemicals may limit the growth of established plants or the germination and growth of seeds and seedlings.

Another of the ways in which the ecological role of a plant can make it a weed even if it is in itself inoffensive, is if it harbours a pest that is dependent on it for survival; for example, *Berberis* species are intermediate hosts for stem rust fungi, so that they promote serious damage to wheat crops when growing near the fields.

Competition with Cultivated and Endemic Plants

700 cattle that were killed overnight by a poisonous weed.

A number of native or non-native plants are unwanted in a specific location for a number of reasons. An important one is that they interfere with food and fiber production in agriculture, wherein they must be controlled in order to prevent lost or diminished crop yields. Other important reasons are that they interfere with other cosmetic, decorative, or recreational goals, such as in lawns, landscape architecture, playing fields, and golf courses. Similarly, they can be of concern for environmental reasons whereby introduced species out-compete for resources or space with desired endemic plants. For all these reasons; horticulture, both functional and cosmetic, and environmental, weeds interfere by:

- Competing with the desired plants for the resources that a plant typically needs, namely, direct sunlight, soil nutrients, water, and (to a lesser extent) space for growth;

- Providing hosts and vectors for plant pathogens, giving them greater opportunity to infect and degrade the quality of the desired plants;

- Providing food or shelter for animal pests such as seed-eating birds and tephritid fruit flies that otherwise could hardly survive seasonal shortages;

- Offering irritation to the skin or digestive tracts of people or animals, either physical irritation via thorns, prickles, or burs, or chemical irritation via natural poisons or irritants in the weed (for example, the poisons found in *nerium* species);

- Causing root damage to engineering works such as drains, road surfaces, and foundations, blocking streams and rivulets.

In weed ecology some authorities speak of the relationship between "the three Ps": plant, place, perception. These have been very variously defined, but the weed traits listed by H.G. Baker are widely cited.

Weeds have long been a concern, perhaps as long as humans have cultivated plants. They are mentioned in various historic texts, such as a Shakespearean sonnet:

"To thy fair flower add the rank smell of weeds: But why thy odour matcheth not thy show. The soil is this, that thou dost common grow."

and the Bible:

"Cursed is the ground because of you; through painful toil you will eat of it all the days of your life. It will produce thorns and thistles for you, and you will eat the plants of the field. By the sweat of your brow you will eat your food until you return to the ground."

Benefits of Weed Species

While the term "weed" generally has a negative connotation, many plants known as weeds can have beneficial properties. A number of weeds, such as the dandelion (*Taraxacum*) and lamb's quarter, are edible, and their leaves or roots may be used for food or herbal medicine. Burdock is common over much of the world, and is sometimes used to make soup and medicine in East Asia. Some weeds attract beneficial insects, which in turn can protect crops from harmful pests. Weeds can also prevent pest insects from finding a crop, because their presence disrupts the incidence of positive cues which pests use to locate their food. Weeds may also act as a "living mulch", providing ground cover that reduces moisture loss and prevents erosion. Weeds may also improve soil fertility; dandelions, for example, bring up nutrients like calcium and nitrogen from deep in the soil with their tap root, and clover hosts nitrogen-fixing bacteria in its roots, fertilizing the soil directly. The dandelion is also one of several species which break up hardpan in overly cultivated fields, helping crops grow deeper root systems. Some garden flowers originated as weeds in cultivated fields and have been selectively bred for their garden-worthy flowers or foliage. An example of a crop weed that is grown in gardens is the corncockle, (*Agrostemma githago*), which was a common weed in European wheat fields, but is now sometimes grown as a garden plant.

Dispersal

Many weed species have moved out of their natural geographic ranges and spread around the world in tandem with human migrations and commerce. Weed seeds are often collected and transported with crops after the harvesting of grains, so humans are a vector of transport as well as a producer of the disturbed environments to which weed species are well adapted, resulting in many weeds having a close association with human activities.

Some weed species have been classified as noxious weeds by government authorities because, if left unchecked, they often compete with native or crop plants or cause harm to livestock. They are often foreign species accidentally or imprudently imported into a region where there are few natural controls to limit their population and spread.

Weeds as Adaptable Species

An alternate definition often used by biologists is any species, not just plants, that can quickly adapt to any environment. Some traits of weedy species are the ability to reproduce quickly, disperse widely, live in a variety of habitats, establish a population in strange places, succeed in disturbed ecosystems and resist eradication once established. Such species often do well in human-dominated environments as other species are not able to adapt. Common examples include the common pigeon, brown rat and the raccoon. Other weedy species have been able to expand their range without actually living in human environments, as human activity has damaged the ecosystems of other species. These include the coyote, the white-tailed deer and the brown headed cowbird.

In response to the idea that humans may face extinction due to environmental degradation, paleontologist David Jablonsky counters by arguing that humans are a weed species. Like other weedy species, humans are widely dispersed in a wide variety of environments, and are highly unlikely to go extinct no matter how much damage the environment faces.

Plants often Considered to be Weeds

White clover

White clover is considered by some to be a weed in lawns, but in many other situations is a desirable source of fodder, honey and soil nitrogen.

A short list of some plants that often are considered to be weeds follows:

- Amaranth – ("pigweed") annual with copious long-lasting seeds, also a highly edible and resilient food source,

- Bermuda grass – perennial, spreading by runners, rhizomes and seeds,

- Bindweed,

- Broadleaf plantain – perennial, spreads by seeds that persist in the soil for many years,

- Burdock – biennial,

- Common lambsquarters – annual,

- Creeping charlie – perennial, fast-spreading plants with long creeping stems,

- Dandelion – perennial, wind-spread, fast-growing, and drought-tolerant,

- Goldenrod – perennial,
- Japanese knotweed,
- Kudzu – perennial,
- Leafy spurge – perennial, with underground stems,
- Milk thistle – annual or biennial,
- Poison ivy – perennial,
- Ragweed – annual,
- Sorrel – annual or perennial,
- Striga,
- St John's wort – perennial,
- Sumac – woody perennial,
- Tree of heaven – woody perennial,
- Wild carrot – biennial,
- Wood sorrel – perennial,
- Yellow nutsedge – perennial.

Many invasive weeds were introduced deliberately in the first place, and may have not been considered nuisances at the time, but rather beneficial.

Weed Control

Weed control is the botanical component of pest control, which attempts to stop weeds, especially noxious weeds, from competing with desired flora and fauna including domesticated plants and livestock, and in natural settings preventing non native species competing with native species.

Weed control is important in agriculture. Methods include hand cultivation with hoes, powered cultivation with cultivators, smothering with mulch, lethal wilting with high heat, burning, and chemical control with herbicides (weed killers).

The definition of a weed is context-dependent, and can be taken to mean any plant not wanted for any reason.

Weeds compete with productive crops or pasture, ultimately converting productive land into unusable scrub. Weeds can be poisonous, distasteful, produce burrs, thorns or otherwise interfere with the use and management of desirable plants by contaminating harvests or interfering with livestock.

Weeds compete with crops for space, nutrients, water and light. Smaller, slower growing seedlings are more susceptible than those that are larger and more vigorous. Onions are one of the most vulnerable, because they are slow to germinate and produce slender, upright stems. By contrast broad beans produce large seedlings and suffer far fewer effects other than during periods of water shortage at the crucial time when the pods are filling out. Transplanted crops raised in sterile soil or potting compost gain a head start over germinating weeds.

Weeds also vary in their competitive abilities and according to conditions and season. Tall-growing vigorous weeds such as fat hen (*Chenopodium album*) can have the most pronounced effects on adjacent crops, although seedlings of fat hen that appear in late summer produce only small plants. Chickweed (*Stellaria media*), a low growing plant, can happily co-exist with a tall crop during the summer, but plants that have overwintered will grow rapidly in early spring and may swamp crops such as onions or spring greens.

The presence of weeds does not necessarily mean that they are damaging a crop, especially during the early growth stages when both weeds and crops can grow without interference. However, as growth proceeds they each begin to require greater amounts of water and nutrients. Estimates suggest that weed and crop can co-exist harmoniously for around three weeks before competition becomes significant. One study found that after competition had started, the final yield of onion bulbs was reduced at almost 4% per day.

Perennial weeds with bulbils, such as lesser celandine and oxalis, or with persistent underground stems such as couch grass (*Agropyron repens*) or creeping buttercup (*Ranunculus repens*) store reserves of food, and are thus able to grow faster and with more vigour than their annual counterparts. Some perennials such as couch grass exude allelopathic chemicals that inhibit the growth of other nearby plants.

Weeds can also host pests and diseases that can spread to cultivated crops. Charlock and Shepherd's purse may carry clubroot, eelworm can be harboured by chickweed, fat hen and shepherd's purse, while the cucumber mosaic virus, which can devastate the cucurbit family, is carried by a range of different weeds including chickweed and groundsel.

Insect pests often do not attack weeds. However pests such as cutworms may first attack weeds then move on to cultivated crops.

Some plants are considered weeds by some farmers and crops by others. Charlock, a common weed in the southeastern US, are weeds according to row crop growers, but are valued by beekeepers, who seek out places where it blooms all winter, thus providing pollen for honeybees and other pollinators. Its bloom resists all but a very hard freeze, and recovers once the freeze ends.

Weed Propagation

Seeds

Annual and biennial weeds such as chickweed, annual meadow grass, shepherd's purse, groundsel, fat hen, cleaver, speedwell and hairy bittercress propagate themselves by seeding. Many produce huge numbers of seed several times a season, some all year round. Groundsel can produce 1000 seed, and can continue right through a mild winter, whilst Scentless Mayweed produces over 30,000 seeds per plant. Not all of these will germinate at once, but over several seasons, lying dormant in the soil sometimes for years until exposed to light. Poppy seed can survive 80–100 years, dock 50 or more. There can be many thousands of seeds in a square foot or square metre of ground, thus any soil disturbance will produce a flush of fresh weed seedlings.

Subsurface/Surface

The most persistent perennials spread by underground creeping rhizomes that can re-grow from a tiny fragment. These include couch grass, bindweed, ground elder, nettles, rosebay willow herb, Japanese knotweed, horsetail and bracken, as well as creeping thistle, whose tap roots can put out lateral roots. Other perennials put out runners that spread along the soil surface. As they creep they set down roots, enabling them to colonise bare ground with great rapidity. These include creeping buttercup and ground ivy. Yet another group of perennials propagate by stolons- stems that arch back into the ground to reroot. The most familiar of these is the bramble.

Methods

Weed control plans typically consist of many methods which are divided into biological, chemical, cultural, and physical/mechanical control.

Pesticide-free thermic weed control with a weed burner on a potato field.

Physical/Mechanical Methods

Coverings

In domestic gardens, methods of weed control include covering an area of ground with a material that creates a hostile environment for weed growth, known as a *weed mat*.

Several layers of wet newspaper prevent light from reaching plants beneath, which kills them. Daily saturating the newspaper with water plant decomposition. After several weeks, all germinating weed seeds are dead.

In the case of black plastic, the greenhouse effect kills the plants. Although the black plastic sheet is effective at preventing weeds that it covers, it is difficult to achieve complete coverage. Eradicating persistent perennials may require the sheets to be left in place for at least two seasons.

Some plants are said to produce root exudates that suppress herbaceous weeds. *Tagetes minuta* is claimed to be effective against couch and ground elder, whilst a border of comfrey is also said to act as a barrier against the invasion of some weeds including couch. A 5–10 centimetres (2.0–3.9 in) layer of wood chip mulch prevents most weeds from sprouting. Gravel can serve as an inorganic mulch.

Tools used for amateur weeding include spades and gloves.

Irrigation is sometimes used as a weed control measure such as in the case of paddy fields to kill any plant other than the water-tolerant rice crop.

Manual Removal

Many gardeners still remove weeds by manually pulling them out of the ground, making sure to include the roots that would otherwise allow them to resprout.

Hoeing off weed leaves and stems as soon as they appear can eventually weaken and kill perennials, although this will require persistence in the case of plants such as bindweed. Nettle infestations can be tackled by cutting back at least three times a year, repeated over a three-year period. Bramble can be dealt with in a similar way.

Weeds are removed manually in large parts of India.

A highly successful, mostly manual, removal programme of weed control in natural bush land has been the control of sea spurge by Sea Spurge Remote Area Teams in Tasmania.

Ploughing includes tilling of soil, intercultural ploughing and summer ploughing. Ploughing uproots weeds, causing them to die. In summer ploughing is done during deep summers. Summer ploughing also helps in killing pests.

Mechanical tilling can remove weeds around crop plants at various points in the growing process.

An Aquamog can be used to remove weeds covering a body of water.

Thermal

Several thermal methods can control weeds.

Flame weeders use a flame several centimeters away from the weeds to give them a sudden and severe heating. The goal of flame weeding is not necessarily burning the plant, but rather causing a lethal wilting by denaturing proteins in the weed. Similarly, hot air weeders can heat up the seeds to the point of destroying them. Flame weeders can be combined with techniques such as stale seedbeds (preparing and watering the seedbed early, then killing the nascent crop of weeds that springs up from it, then sowing the crop seeds) and preemergence flaming (doing a flame pass against weed seedlings after the sowing of the crop seeds but before those seedlings emerge from the soil—a span of time that can be days or weeks).

Hot foam (foamstream) causes the cell walls to rupture, killing the plant. Weed burners heat up soil quickly and destroy superficial parts of the plants. Weed seeds are often heat resistant and even react with an increase of growth on dry heat.

Since the 19th century soil steam sterilization has been used to clean weeds completely from soil. Several research results confirm the high effectiveness of humid heat against weeds and its seeds.

Soil solarization in some circumstances is very effective at eliminating weeds while maintaining grass. Planted grass tends to have a higher heat/humidity tolerance than unwanted weeds.

Seed Targeting

In 1998, the Australian Herbicide Resistance Initiative (AHRI), debuted. Gathered fifteen scientists and technical staff members to conduct field surveys, collect seeds, test for resistance and study the biochemical and genetic mechanisms of resistance. A collaboration with DuPont led to a mandatory herbicide labeling program, in which each mode of action is clearly identified by a letter of the alphabet.

The key innovation of the AHRI approach has been to focus on weed seeds. Ryegrass seeds last only a few years in soil, so if farmers can prevent new seeds from arriving, the number of sprouts will shrink each year. Until the new approach farmers were unintentionally helping the seeds. Their combines loosen ryegrass seeds from their stalks and spread them over the fields. In the mid-1980s, a few farmers hitched covered trailers, called "chaff carts", behind their combines to catch the chaff and weed seeds. The collected material is then burned.

An alternative is to concentrate the seeds into a half-meter-wide strip called a windrow and burn the windrows after the harvest, destroying the seeds. Since 2003, windrow burning has been adopted by about 70% of farmers in Western Australia.

Yet another approach is the Harrington Seed Destructor, which is an adaptation of a coal pulverizing cage mill that uses steel bars whirling at up to 1500 rpm. It keeps all the organic material in the field and does not involve combustion, but kills 95% of seeds.

Cultural Methods

Stale Seed Bed

Another manual technique is the 'stale seed bed', which involves cultivating the soil, then leaving it fallow for a week or so. When the initial weeds sprout, the grower lightly hoes them away before planting the desired crop. However, even a freshly cleared bed is susceptible to airborne seed from elsewhere, as well as seed carried by passing animals on their fur, or from imported manure.

Buried Drip Irrigation

Buried drip irrigation involves burying drip tape in the subsurface near the planting bed, thereby limiting weeds access to water while also allowing crops to obtain moisture. It is most effective during dry periods.

Crop Rotation

Rotating crops with ones that kill weeds by choking them out, such as hemp, Mucuna pruriens, and other crops, can be a very effective method of weed control. It is a way to avoid the use of herbicides, and to gain the benefits of crop rotation.

Biological Methods

A biological weed control regiment can consist of biological control agents, bioherbicides, use of grazing animals, and protection of natural predators. Post-dispersal, weed seed predators, like ground beetles and small vertebrates, can substantially contribute to the weed regulation by removing weed seeds from the soil surface and thus reduce seed bank size. Several studies provided evidence for the role of invertebrates to the biological control of weeds.

Animal Grazing

Companies using goats to control and eradicate leafy spurge, knapweed, and other toxic weeds have sprouted across the American West.

Chemical Methods

Organic Approaches

Weed control, circa 1930-40s.

A mechanical weed control device: the diagonal weeder.

Organic weed control involves anything other than applying manufactured chemicals. Typically a combination of methods are used to achieve satisfactory control.

Sulfur in some circumstances is accepted within British Soil Association standards.

Herbicides

The above described methods of weed control use no or very limited chemical inputs. They are preferred by organic gardeners or organic farmers.

However weed control can also be achieved by the use of herbicides. Selective herbicides kill certain targets while leaving the desired crop relatively unharmed. Some of these act by interfering with the growth of the weed and are often based on plant hormones. Herbicides are generally classified as follows:

- Contact herbicides destroy only plant tissue that contacts the herbicide. Generally, these are the fastest-acting herbicides. They are ineffective on perennial plants that can re-grow from roots or tubers.

- Systemic herbicides are foliar-applied and move through the plant where they destroy a greater amount of tissue. Glyphosate is currently the most used systemic herbicide.

- Soil-borne herbicides are applied to the soil and are taken up by the roots of the target plant.

- Pre-emergent herbicides are applied to the soil and prevent germination or early growth of weed seeds.

In agriculture large scale and systematic procedures are usually required, often by machines, such as large liquid herbicide 'floater' sprayers, or aerial application.

Bradley Method

Bradley Method of Bush Regeneration uses ecological processes to do much of the work. Perennial weeds also propagate by seeding; the airborne seed of the dandelion and the rose-bay willow herb parachute far and wide. Dandelion and dock also put down deep tap roots, which, although they do not spread underground, are able to regrow from any remaining piece left in the ground.

Hybrid

One method of maintaining the effectiveness of individual strategies is to combine them with others that work in complete different ways. Thus seed targeting has been combined with herbicides. In Australia seed management has been effectively combined with trifluralin and clethodim.

Resistance

Resistance occurs when a target adapts to circumvent a particular control strategy. It affects not only weed control,but antibiotics, insect control and other domains. In agriculture is mostly considered in reference to pesticides, but can defeat other strategies, e.g., when a target species becomes more drought tolerant via selection pressure.

Farming Practices

Herbicide resistance recently became a critical problem as many Australian sheep farmers switched to exclusively growing wheat in their pastures in the 1970s. In wheat fields, introduced varieties of ryegrass, while good for grazing sheep, are intense competitors with wheat. Ryegrasses produce so many seeds that, if left unchecked, they can completely choke a field. Herbicides provided excellent control, while reducing soil disrupting because of less need to plough. Within little more than a decade, ryegrass and other weeds began to develop resistance. Australian farmers evolved again and began diversifying their techniques.

In 1983, patches of ryegrass had become immune to Hoegrass, a family of herbicides that inhibit an enzyme called acetyl coenzyme A carboxylase.

Ryegrass populations were large, and had substantial genetic diversity, because farmers had planted many varieties. Ryegrass is cross-pollinated by wind, so genes shuffle frequently. Farmers sprayed inexpensive Hoegrass year after year, creating selection pressure, but were diluting the herbicide in order to save money, increasing plants survival. Hoegrass was mostly replaced by a group of herbicides that block acetolactate synthase, again helped by poor application practices. Ryegrass evolved a kind of "cross-resistance" that allowed it to rapidly break down a variety of herbicides. Australian farmers lost four classes of herbicides in only a few years. As of 2013 only two herbicide classes, called Photosystem II and long-chain fatty acid inhibitors, had become the last hope.

Strategic Weed Management

Strategic weed management is a process of managing weeds at a district, regional or national scale. In Australia the first published weed management strategies were developed in Tasmania, New South Wales and South Australian 1999, followed by the National Weeds Strategy in 1999.

Internationally weed societies help link weed science and management. In North America the Weed Science Society of America (WSSA) is the major weed science learned body with European Weed Research Council and Council of Australasian Weed Society serving the same purpose in Australia and New Zealand. There are also regional weed societies in Australasia and North America.

References

- Weed, plant: britannica.com, Retrieved 13 March, 2019

- Bell, graham (2005). The permaculture garden. Chelsea green publishing. Pp. 63–64. Isbn 9781856230278

- Hans lambers; f stuart chapin iii; thijs l. Pons (8 october 2008). Plant physiological ecology. Springer. Pp. 507–. Isbn 978-0-387-78341-3

- Bohan d.; et al. (2011). "national-scale regulation of the weed seedbank by carabid predators". Journal of applied ecology. 48(4): 888–898

- Cloutier, daniel. "european weed research society (ewrs) - home page". Www.ewrs.org. Retrieved 2018-09-11

6
Pest Control

The organisms that reduce the quality, availability and value of human resources are classified as pests. Insects, mites, rodents, molluscs and birds are the various types of pests. The following chapter elucidates the different techniques and methods associated with pest control such as physical pest control, fumigation, biological pest control, etc.

Pest

A pest species can be any species that humans consider undesirable. Any organism that reduces the availability, quality, or value of a human resource can be classified as a pest. This designation in no way reflects the organism's role in the natural ecosystem but is more an indicator that they are in conflict with humans. Plant pests, also referred to as weeds.

A pest in one area may not be considered a pest elsewhere. Often organisms rise to pest status because they escape normal control by natural regulating agents. This is achieved through direct or indirect importation to a new region or by human activities which reduce or eliminate the efficiency of their natural enemies. Without controls on population growth, organisms can rapidly achieve levels at which damage is caused thus becoming pests (e.g., locust swarms stripping landscapes bare). However, organisms do not need to exist in large numbers to be a pest. For example, the codling moth (Cydia pomonella) does not lay many eggs compared to many insects and often produces only one generation each year. However, because it blemishes apples, making them commercially undesirable, the codling moth is considered an important agricultural pest.

Pest management is therefore a means to reduce pest numbers to an acceptable threshold. An acceptable threshold, in most cases, refers to an economically justifiable threshold where application of pest control measures reduces pest numbers to a level below which additional applications would not be profitable (i.e., where additional costs of control exceed additional benefits). Pest eradication (i.e., complete removal) is usually not a viable option.

Types of Pests

Ants, snails, slugs, caterpillars, spider mites, whiteflies, fungus gnats, and aphids are the most common types of garden pests. It is an organism that causes any type of damage to the crop plants.

- Insect: These are important and major pests. Insects have three pairs of legs, two pairs of wings, segmented body and characteristic compound eyes and antennae. Insects are tricky and cause damage in different ways viz. sucking sap from plants, biting plant parts, boring in to fruits, twigs and leaves, attacking roots, barks and blossoms etc. The damaging stages of different insect pests are larvae, adults and nymphs.

- Mites: These are creatures like insect but have soft body and four pairs of legs. These tiny creatures have red or pale yellow colour. They suck the sap from the plant and attack the crops in huge number.

- Rodents: This group of pest eat away large amount of human food and also damage the crops on large scale. They are also responsible for heavy loss to stored grains on farms, in warehouses and houses.

- Animals: Animals like Wild Boar, Deer, Elephants, Wild Buffalo, Jackals, Monkeys, Squirrels cause direct damage to crop plants. They eat away the plants and by and large they waste huge amount of crops.

- Birds: Birds attack the crop plants and eat grains. Crow, Parrots and Sparrows are major among birds that attack the crops.

Pest Control

Pest control is the regulation or management of a species defined as a pest, a member of the animal kingdom that impacts adversely on human activities. The human response depends on the importance of the damage done, and will range from tolerance, through deterrence and management, to attempts to completely eradicate the pest. Pest control measures may be performed as part of an integrated pest management strategy.

In agriculture, pests are kept at bay by cultural, chemical and biological means. Ploughing and cultivation of the soil before sowing reduces the pest burden and there is a modern trend to limit the use of pesticides as far as possible. This can be achieved by monitoring the crop, only applying insecticides when necessary, and by growing varieties and crops which are resistant to pests. Where possible, biological means are used, encouraging the natural enemies of the pests and introducing suitable predators or parasites.

In homes and urban environments, the pests are the rodents, birds, insects and other organisms that share the habitat with humans, and that feed on and spoil possessions. Control of these pests is attempted through exclusion, repulsion, physical removal or chemical means. Alternatively, various methods of biological control can be used including sterilisation programmes.

Bronze cat, Ancient Egypt.

Pest control is at least as old as agriculture, as there has always been a need to keep crops free from pests. As long ago as 3000 BC in Egypt, cats were used to control pests of grain stores such as rodents. Ferrets were domesticated by 500 AD in Europe for use as mousers. Mongooses were introduced into homes to control rodents and snakes, probably by the ancient Egyptians.

The conventional approach was probably the first to be employed, since it is comparatively easy to destroy weeds by burning them or ploughing them under, and to kill larger competing herbivores. Techniques such as crop rotation, companion planting (also known as intercropping or mixed cropping), and the selective breeding of pest-resistant cultivars have a long history.

Red weaver ants, here feeding on a snail, have been used to control pests in China, Southeast Asia and Africa for many centuries.

Chemical pesticides were first used around 2500 BC, when the Sumerians used sulphur compounds as insecticides. Modern pest control was stimulated by the spread across the United States of the Colorado potato beetle. After much discussion, arsenical compounds were used to control the beetle and the predicted poisoning of the human population did not occur. This led the way to a widespread acceptance of insecticides across the continent. With the industrialisation and mechanization of agriculture in the 18th and 19th centuries, and the introduction of the insecticides pyrethrum and derris, chemical pest control became widespread. In the 20th century, the discovery of several synthetic insecticides, such as DDT, and herbicides boosted this development.

Biological control is first recorded around 300 AD in China, when colonies of weaver ants, *Oecophylla smaragdina*, were intentionally placed in citrus plantations to control beetles and caterpillars. Also in China, ducks were used in paddy fields to consume pests, as illustrated in ancient cave art. In 1762, an Indian mynah was brought to Mauritius to control locusts, and about the same time, citrus trees in Burma were connected by bamboos to allow ants to pass between them and help control caterpillars. In the 1880s, ladybirds were used in citrus plantations in California to control scale insects, and other biological control experiments followed. The introduction of DDT, a cheap and effective compound, put an effective stop to biological control experiments. By the 1960s, problems of resistance to chemicals and damage to the environment began to emerge, and biological control had a renaissance. Chemical pest control is still the predominant type of pest control today, although a renewed interest in traditional and biological pest control developed towards the end of the 20th century and continues to this day.

In Agriculture, Horticulture and Forestry

Control Methods - Biological Pest Control

Biological pest control: parasitoid wasp (Cotesia congregata) adult with pupal cocoons on its host, a tobacco hornworm Manduca sexta (green background).

Biological pest control is a method of controlling pests such as insects and mites by using other organisms. It relies on predation, parasitism, herbivory or other natural mechanisms, but typically also involves an active human management role. Classical biological control involves the introduction of natural enemies of the pest that are bred in the laboratory and released into the environment. An alternative approach is to augment the natural enemies that occur in a particular area by releasing more, either in small, repeated batches, or in a single large-scale release. Ideally, the released organism will breed and survive, and provide long-term control. Biological control can be an important component of an integrated pest management programme.

For example: mosquitoes are often controlled by putting *Bt Bacillus thuringiensis* ssp. *israelensis*, a bacterium that infects and kills mosquito larvae, in local water sources.

Cultural Control

Cultivation by ploughing exposes insect pests to predators
such as black-headed gulls.

Mechanical pest control is the use of hands-on techniques as well as simple equipment and devices, that provides a protective barrier between plants and insects. This is referred to as tillage and is one of the oldest methods of weed control as well as being useful for pest control; wireworms, the larvae of the common click beetle, are very destructive pests of newly ploughed grassland, and repeated cultivation exposes them to the birds and other predators that feed on them.

Crop rotation can help to control pests by depriving them of their host plants. It is a major tactic in the control of corn rootworm, and has reduced early season incidence of Colorado potato beetle by as much as 95%.

Trap Cropping

A trap crop is a crop of a plant that attracts pests, diverting them from nearby crops. Pests aggregated on the trap crop can be more easily controlled using pesticides or other methods. However, trap-cropping, on its own, has often failed to cost effectively reduce pest densities on large commercial scales, without the use of pesticides, possibly due to the pests' ability to disperse back into the main field.

Pesticides

Pesticides are applied to crops by agricultural aircraft, tractor-mounted crop sprayers or as seed dressings to control pests. However, successful control by pesticides is not easy; the right formulation must be chosen, the timing is often critical, the method of application is important, adequate coverage and retention on the crop are necessary. The killing of natural enemies of the target pest should be minimised. This is particularly important in countries where there are natural reservoirs of pests and their enemies in the countryside surrounding plantation crops, and these co-exist in a delicate balance. Often in less-developed countries, the crops are well adapted to the local situation and no pesticides are needed. Where progressive farmers are using fertilisers to grow improved crop varieties, these are often more susceptible to pest damage, but the indiscriminate application of pesticides may be detrimental in the longer term.

Spraying pine logs with insecticide against Ips
sexdentatus, a pine engraver beetle.

The efficacy of chemical pesticides tends to diminish over time. This is because any organism that manages to survive the initial application will pass on its genes to its offspring and a resistant strain will be developed. In this way, some of the most serious pests have developed resistance and are no longer killed by pesticides that used to kill their ancestors. This necessitates higher concentrations of chemical, more frequent applications and a movement to more expensive formulations.

Pesticides are formulated to kill pests, but many have detrimental effects on non-target species; of particular concern is the damage done to honey-bees, solitary bees and other pollinating insects and in this regard, the time of day when the spray is applied can be important. The widely used neonicotinoids have been banned on flowering crops in some countries because of their effects on bees. Some pesticides may cause cancer and other health problems in humans, as well as being harmful to wildlife. There can be acute effects immediately after exposure or chronic effects after continuous low-level, or occasional exposure. Maximum residue limits for pesticides in foodstuffs and animal feed are set by many nations.

Spruce budworm (adult and pupa shown), a serious pest of forests, can be
monitored using pheromone traps.

Forestry

Forest pests present a significant problem because it is not easy to access the canopy and monitor pest populations. In addition, forestry pests such as bark beetles, kept under control by natural enemies in their native range, may be transported large distances in cut timber to places where they have no natural predators, enabling them to cause extensive economic damage. Pheromone traps have been used to monitor pest populations in the canopy. These release volatile chemicals that attract males. Pheromone traps can detect the arrival of pests or alert foresters to outbreaks. For example, the spruce budworm, a destructive pest of spruce and balsam fir, has been monitored using pheromone traps in Canadian forests for several decades. In some regions, such as New Brunswick, areas of forest are sprayed with pesticide to control the budworm population and prevent the damage caused during outbreaks.

In Homes and Cities

Many unwelcome animals visit or make their home in residential buildings, industrial sites and urban areas. Some contaminate foodstuffs, damage structural timbers, chew through fabrics or infest stored dry goods. Some inflict great economic loss, others carry diseases or cause fire hazards, and some are just a nuisance. Control of these pests has been attempted by improving sanitation and garbage control, modifying the habitat, and using repellents, growth regulators, traps, baits and pesticides.

General Methods

Physical Pest Control

Dog control van.

Physical pest control involves trapping or killing pests such as insects and rodents. Historically, local people or paid rat-catchers caught and killed rodents using dogs and traps. On a domestic scale, sticky flypapers are used to trap flies. In larger buildings, insects may be trapped using such means as pheromones, synthetic volatile chemicals or ultraviolet light to attract the insects; some have a sticky base or an electrically charged grid to kill them. Glueboards are sometimes used for monitoring cockroaches and to catch rodents. Rodents can be killed by suitably baited spring traps and can be caught in cage traps for relocation. Talcum powder or "tracking powder" can be used to establish routes used by rodents inside buildings and acoustic devices can be used for detecting beetles in structural timbers.

Historically, firearms have been one of the primary methods used for pest control. "Garden Guns" are smooth bore shotguns specifically made to fire .22 caliber snake shot or 9mm Flobert, and are commonly used by gardeners and farmers for snakes, rodents, birds, and other pest. Garden Guns are short range weapons that can do little harm past 15 to 20 yards, and they're relatively quiet when fired with snake shot, compared to a standard ammunition. These guns are especially effective inside of barns and sheds, as the snake shot will not shoot holes in the roof or walls, or more importantly injure livestock with a ricochet. They are also used for pest control at airports, warehouses, stockyards, etc.

The most common shot cartridge is .22 Long Rifle loaded with #12 shot. At a distance of about 10 feet (3 m), which is about the maximum effective range, the pattern is about 8 inches (20 cm) in diameter from a standard rifle. Special smoothbore shotguns, such as the Marlin Model 25MG can produce effective patterns out to 15 or 20 yards using .22 WMR shotshells, which hold 1/8 oz. of #12 shot contained in a plastic capsule.

Poisoned Bait

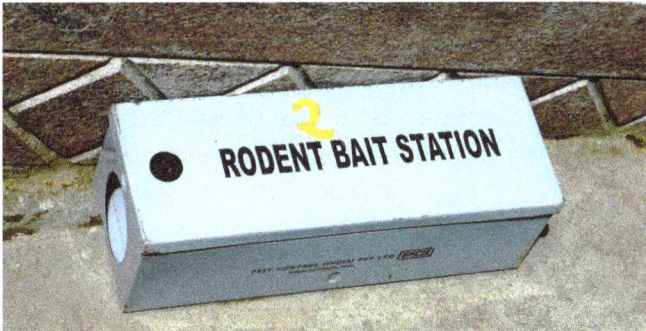

Rodent bait station.

Poisoned bait is a common method for controlling rats, mice, birds, slugs, snails, ants, cockroaches and other pests. The basic granules, or other formulation, contains a food attractant for the target species and a suitable poison. For ants, a slow-acting toxin is needed so that the workers have time to carry the substance back to the colony, and for flies, a quick-acting substance to prevent further egg-laying and nuisance. Baits for slugs and snails often contain the molluscide metaldehyde, dangerous to children and household pets.

Warfarin has traditionally been used to kill rodents, but many populations have developed resistance to this anticoagulant, and difenacoum is often substituted. These are cumulative poisons, requiring bait stations to be topped up regularly. Poisoned meat has been used for centuries to kill animals such as wolves and birds of prey. Poisoned carcasses however kill a wide range of carrion feeders, not only the targeted species. Raptors in Israel were nearly wiped out following a period of intense poisoning of rats and other crop pests.

Fumigation

Tent fumigation of a house.

Fumigation is the treatment of a structure to kill pests such as wood-boring beetles by sealing it or surrounding it with an airtight cover such as a tent, and fogging with liquid insecticide for an extended period, typically of 24–72 hours. This is costly and inconvenient as the structure cannot be used during the treatment, but it targets all life stages of pests.

An alternative, space treatment, is fogging or misting to disperse a liquid insecticide in the atmosphere within a building without evacuation or airtight sealing, allowing most work within the building to continue, at the cost of reduced penetration. Contact insecticides are generally used to minimise long lasting residual effects.

Sterilization

Populations of pest insects can sometimes be dramatically reduced by the release of sterile individuals. This involves the mass rearing of a pest, sterilising it by means of X-rays or some other means, and releasing it into a wild population. It is particularly useful where a female only mates once and where the insect does not disperse widely. This technique has been successfully used against the New World screw-worm fly, some species of tsetse fly, tropical fruit flies, the pink bollworm and the codling moth, among others.

Laboratory studies conducted with U-5897 (3-chloro-1,2-propanediol) were attempted in the early 1970s for rat control, although these proved unsuccessful. In 2013, New York City tested sterilization traps, demonstrating a 43% reduction in rat populations. The product ContraPest was approved for the sterilization of rodents by the United States Environmental Protection Agency in August 2016.

Methods for Specific Pests

Natural Rodent Control

Brown rat infestation.

Several wildlife rehabilitation organizations encourage natural form of rodent control through exclusion and predator support and preventing secondary poisoning altogether. The United States Environmental Protection Agency notes in its Proposed Risk Mitigation Decision for Nine Rodenticides that "without habitat modification to make areas less attractive to commensal rodents, even eradication will not prevent new populations from recolonizing the habitat." The United States Environmental Protection Agency has prescribed guidelines for natural rodent control and for safe trapping in residential areas with subsequent release to the wild. People sometimes attempt to limit rodent damage using repellents. Balsam fir oil from the tree Abies balsamea is an EPA approved non-toxic rodent repellent. Acacia polyacantha subsp. Campylacantha root emits chemical compounds that repel animals including rats.

Pantry Pests

The red flour beetle, Tribolium castaneum, attacks stored grain products worldwide.

Insect pests including the Mediterranean flour moth, the Indian mealmoth, the cigarette beetle, the drugstore beetle, the confused flour beetle, the red flour beetle, the merchant grain beetle, the sawtoothed grain beetle, the wheat weevil, the maize weevil and the rice weevil infest stored dry foods such as flour, cereals and pasta.

In the home, foodstuffs found to be infested are usually discarded, and storing such products in sealed containers should prevent the problem from reoccurring. The eggs of these insects are likely to go unnoticed, with the larvae being the destructive life stage, and the adult the most noticeable stage. Since pesticides are not safe to use near food, alternative treatments such as freezing for four days at 0 °F (−18 °C) or baking for half an hour at 130 °F (54 °C) should kill any insects present.

Clothes Moths

The larvae of clothes moths (mainly Tineola bisselliella and Tinea pellionella) feed on fabrics and carpets, particularly those that are stored or soiled. The adult females lay batches of eggs on natural fibres, including wool, silk and fur, as well as cotton and linen in blends. The developing larvae spin protective webbing and chew into the fabric, creating holes and specks of excrement. Damage is often concentrated in concealed locations, under collars and near seams of clothing, in folds and crevices in upholstery and round the edges of carpets as well as under furniture. Methods of control include using airtight containers for storage, periodic laundering of garments, trapping, freezing, heating and the use of chemicals; mothballs contain volatile insect repellents such as 1,4-Dichlorobenzene which deter adults, but to kill the larvae, permethrin, pyrethroids or other insecticides may need to be used.

Larva, pupa and adult clothes moth Tineola bisselliella
with characteristic damage to fabric.

Carpet Beetles

Carpet beetles are members of the family Dermestidae, and while the adult beetles feed on nectar and pollen, the larvae are destructive pests in homes, warehouses and museums. They feed on animal products including wool, silk, leather, fur, the bristles of hair brushes, pet hair, feathers and museum specimens. They tend to infest hidden locations and may feed on larger areas of fabrics than do clothes moths, leaving behind specks of excrement and brown, hollow, bristly-looking cast skins. Management of infestations is difficult and is based on exclusion and sanitation where possible, resorting to pesticides when necessary. The beetles can fly in from outdoors and the larvae can survive on lint fragments, dust and inside the bags of vacuum cleaners. In warehouses and museums, sticky traps baited with suitable pheromones can be used to identify problems, and heating, freezing, spraying the surface with insecticide and fumigation will kill the insects when suitably applied. Susceptible items can be protected from attack by keeping them in clean airtight containers.

Bookworms

Books are sometimes attacked by cockroaches, silverfish, book mites, booklice, and various beetles which feed on the covers, paper, bindings and glue. They leave behind physical damage in the form of tiny holes as well as staining from their faeces. Book pests include the larder beetle, and the larvae of the black carpet beetle and the drugstore beetle which attack leather-bound books, while the common clothes moth and the brown house moth attack cloth bindings. These attacks are largely a problem with historic books, because modern bookbinding materials are less susceptible to this type of damage.

Evidence of attack may be found in the form of tiny piles of book-dust and specks of frass. Damage may be concentrated in the spine, the projecting edges of pages and the cover. Prevention of attack relies on keeping books in cool, clean, dry positions with low humidity, and occasional inspections should be made. Treatment can be by freezing for lengthy periods, but some insect eggs are very resistant and can survive for long periods at low temperatures.

Beetles

House timber split open to reveal larvae of the house longhorn beetle, Hylotrupes bajulus, in their burrows, which are partially filled with frass.

Various beetles in the Bostrichoidea superfamily attack the dry, seasoned wood used as structural timber in houses and to make furniture. In most cases, it is the larvae that do the damage; these are invisible from the outside of the timber, but are chewing away at the wood in the interior of the item. Examples of these are the powderpost beetles, which attack the sapwood of hardwoods, and the furniture beetles, which attacks softwoods, including plywood. The damage has already been done by the time the adult beetles bore their way out, leaving neat round holes behind them. The first that a householder knows about the beetle damage is often when a chair leg breaks off or a piece of structural timber caves in. Prevention is through chemical treatment of the timber prior to its use in construction or in furniture manufacture.

Termites

Termites with colonies in close proximity to houses can extend their galleries underground and make mud tubes to enter homes. The insects keep out of sight and chew their way through structural and decorative timbers, leaving the surface layers intact, as well as through cardboard, plastic and insulation materials. Their presence may become apparent when winged insects appear and swarm in the home in spring. Regular inspection of structures by a trained professional may help detect termite activity before damage becomes substantial. Inspection and monitoring of termites is important because termite alates (winged reproductives) may not always swarm inside a structure. Control and extermination is a professional job involving trying to exclude the insects from the building and trying to kill those already present. Soil-applied liquid termiticides provide a chemical barrier that prevents termites from entering buildings, and lethal baits can be used; these are eaten by foraging insects, and carried back to the nest and shared with other members of the colony, which goes into slow decline.

Mosquitoes

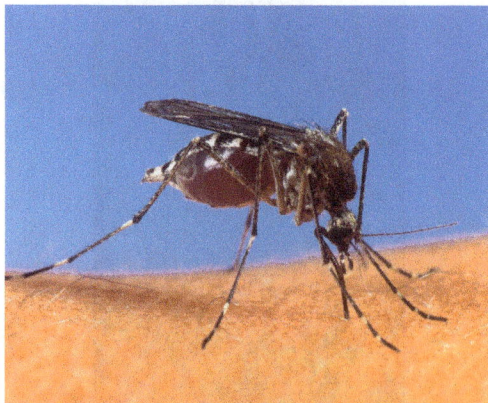

Mosquito (Aedes aegypti) biting a human.

Mosquitoes are midge-like flies in the family Culicidae. Females of most species feed on blood and some act as vectors for malaria and other diseases. Historically they have been controlled by use of DDT and other chemical means, but since the adverse environmental effects of these insecticides has been realised, other means of control have been attempted. The insects rely on water in which to breed and the first line of control is to reduce possible breeding locations by draining marshes and reducing accumulations of standing water. Other approaches include biological control of larvae by the use of fish or other predators, genetic control, the introduction of pathogens, growth-regulating hormones, the release of pheromones and mosquito trapping.

On Airfields

Birds are a significant hazard to aircraft, but it is difficult to keep them away from airfields. Several methods have been explored. Stunning birds by feeding them a bait containing stupefying substances has been tried, and it may be possible to reduce their numbers on airfields by reducing the number of earthworms and other invertebrates by soil treatment. Leaving the grass long on airfields rather than mowing it is also a deterrent to birds. Sonic nets are being trialled; these produce sounds that birds find distracting and seem effective at keeping birds away from affected areas.

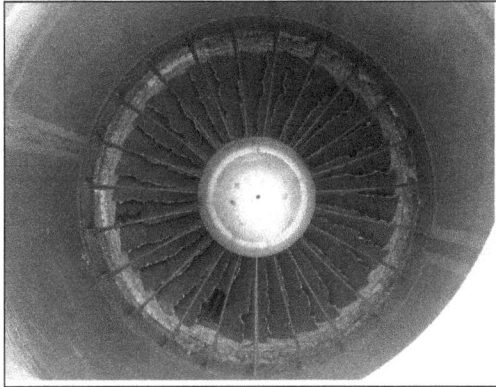

Jet engine fan blades damaged by bird strike.

Biological Pest Control

Biological control or biocontrol is a method of controlling pests such as insects, mites, weeds and plant diseases using other organisms. It relies on predation, parasitism, herbivory, or other natural mechanisms, but typically also involves an active human management role. It can be an important component of integrated pest management (IPM) programs.

There are three basic strategies for biological pest control: classical (importation), where a natural enemy of a pest is introduced in the hope of achieving control; inductive (augmentation), in which a large population of natural enemies are administered for quick pest control; and inoculative (conservation), in which measures are taken to maintain natural enemies through regular reestablishment.

Natural enemies of insect pests, also known as biological control agents, include predators, parasitoids, pathogens, and competitors. Biological control agents of plant diseases are most often referred to as antagonists. Biological control agents of weeds include seed predators, herbivores and plant pathogens.

Biological control can have side-effects on biodiversity through attacks on non-target species by any of the same mechanisms, especially when a species is introduced without thorough understanding of the possible consequences.

Types of Biological Pest Control

There are three basic biological pest control strategies: importation (classical biological control), augmentation and conservation.

Importation

Rodolia cardinalis, the vedalia beetle, was imported from Australia to California in the 19th century, successfully controlling cottony cushion scale.

Importation or classical biological control involves the introduction of a pest's natural enemies to a new locale where they do not occur naturally. Early instances were often unofficial and not based on research, and some introduced species became serious pests themselves.

To be most effective at controlling a pest, a biological control agent requires a colonizing ability which allows it to keep pace with changes to the habitat in space and time. Control is greatest if the agent has temporal persistence, so that it can maintain its population even in the temporary absence of the target species, and if it is an opportunistic forager, enabling it to rapidly exploit a pest population.

Joseph Needham noted a Chinese text dating from 304 AD, Records of the Plants and Trees of the Southern Regions, by Hsi Han, which describes mandarin oranges protected by large reddish-yellow citrus ants which attack and kill insect pests of the orange trees. The citrus ant (Oecophylla smaragdina) was rediscovered in the 20th century, and since 1958 has been used in China to protect orange groves.

One of the earliest successes in the west was in controlling Icerya purchasi (cottony cushion scale) in Australia, using a predatory insect Rodolia cardinalis (the vedalia beetle). This success was repeated in California using the beetle and a parasitoidal fly, Cryptochaetum iceryae. Other successful cases include the control of Antonina graminis in Texas by Neodusmetia sangwani in the 1960s.

Damage from Hypera postica, the alfalfa weevil, a serious introduced pest of forage, was substantially reduced by the introduction of natural enemies. 20 years after their introduction the population of weevils in the alfalfa area treated for alfalfa weevil in the Northeastern United States remained 75 percent down.

Alligator weed was introduced to the United States from South America. It takes root in shallow water, interfering with navigation, irrigation, and flood control. The alligator weed flea beetle and two other biological controls were released in Florida, greatly reducing the amount of land covered by the plant. Another aquatic weed, the giant salvinia (Salvinia molesta) is a serious pest, covering waterways, reducing water flow and harming native species. Control with the salvinia weevil (Cyrtobagous salviniae) and the salvinia stem-borer moth (Samea multiplicalis) is effective in warm climates, and in Zimbabwe, a 99% control of the weed was obtained over a two-year period.

The invasive species Alternanthera philoxeroides (alligator weed) was controlled in Florida (U.S.) by introducing alligator weed flea beetle.

Small commercially reared parasitoidal wasps, Trichogramma ostriniae, provide limited and erratic control of the European corn borer (Ostrinia nubilalis), a serious pest. Careful formulations of the bacterium Bacillus thuringiensis are more effective.

The population of Levuana iridescens, the Levuana moth, a serious coconut pest in Fiji, was brought under control by a classical biological control program in the 1920s.

Augmentation

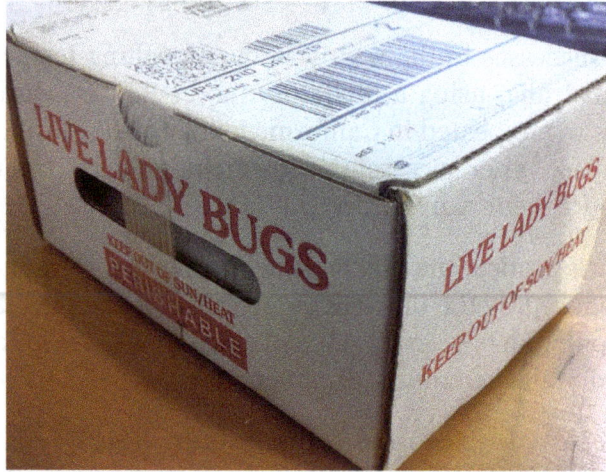

Hippodamia convergens, the convergent lady beetle,
is commonly sold for biological control of aphids.

Augmentation involves the supplemental release of natural enemies that occur in a particular area, boosting the naturally occurring populations there. In inoculative release, small numbers of the control agents are released at intervals to allow them to reproduce, in the hope of setting up longer-term control, and thus keeping the pest down to a low level, constituting prevention rather than cure. In inundative release, in contrast, large numbers are released in the hope of rapidly reducing a damaging pest population, correcting a problem that has already arisen. Augmentation can be effective, but is not guaranteed to work, and depends on the precise details of the interactions between each pest and control agent.

An example of inoculative release occurs in the horticultural production of several crops in greenhouses. Periodic releases of the parasitoidal wasp, *Encarsia formosa*, are used to control greenhouse whitefly, while the predatory mite *Phytoseiulus persimilis* is used for control of the two-spotted spider mite.

The egg parasite Trichogramma is frequently released inundatively to control harmful moths. Similarly, Bacillus thuringiensis and other microbial insecticides are used in large enough quantities for a rapid effect. Recommended release rates for Trichogramma in vegetable or field crops range from 5,000 to 200,000 per acre (1 to 50 per square metre) per week according to the level of pest infestation. Similarly, nematodes that kill insects (that are entomopathogenic) are released at rates of millions and even billions per acre for control of certain soil-dwelling insect pests.

Conservation

The conservation of existing natural enemies in an environment is the third method of biological pest control. Natural enemies are already adapted to the habitat and to the target pest, and their conservation can be simple and cost-effective, as when nec-tar-producing crop plants are grown in the borders of rice fields. These provide nectar to support parasitoids and predators of planthopper pests and have been demonstrated to be so effective (reducing pest densities by 10- or even 100-fold) that farmers sprayed 70% less insecticides and enjoyed yields boosted by 5%. Predators of aphids were sim-ilarly found to be present in tussock grasses by field boundary hedges in England, but they spread too slowly to reach the centres of fields. Control was improved by planting a metre-wide strip of tussock grasses in field centres, enabling aphid predators to over-winter there.

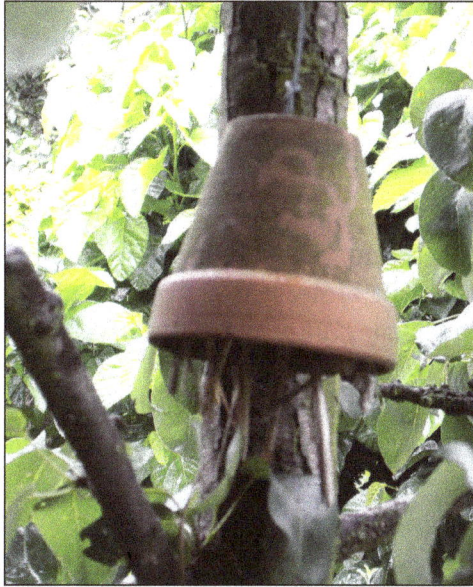

An inverted flowerpot filled with straw to attract earwigs.

Cropping systems can be modified to favor natural enemies, a practice sometimes re-ferred to as habitat manipulation. Providing a suitable habitat, such as a shelterbelt, hedgerow, or beetle bank where beneficial insects such as parasitoidal wasps can live and reproduce, can help ensure the survival of populations of natural enemies. Things as simple as leaving a layer of fallen leaves or mulch in place provides a suitable food source for worms and provides a shelter for insects, in turn being a food source for such beneficial mammals as hedgehogs and shrews. Compost piles and stacks of wood can provide shelter for invertebrates and small mammals. Long grass and ponds support amphibians. Not removing dead annuals and non-hardy plants in the autumn allows insects to make use of their hollow stems during winter. In California, prune trees are sometimes planted in grape vineyards to provide an improved overwintering habitat or refuge for a key grape pest parasitoid. The providing of artificial shelters in the form of wooden caskets, boxes or flowerpots is also sometimes undertaken, particularly in gar-dens, to make a cropped area more attractive to natural enemies. For example, earwigs are natural predators which can be encouraged in gardens by hanging upside-down flowerpots filled with straw or wood wool. Green lacewings can be encouraged by using plastic bottles with an open bottom and a roll of cardboard inside. Birdhouses enable insectivorous birds to nest; the most useful birds can be attracted by choosing an opening just large enough for the desired species.

In cotton production, the replacement of broad-spectrum insecticides with selective control measures such as Bt cotton can create a more favorable environment for natural enemies of cotton pests due to reduced insecticide exposure risk. Such predators or parasitoids can control pests not affected by the Bt protein. Reduced prey quality and abundance associated increased control from Bt cotton can also indirectly decrease natural enemy populations in some cases, but the percentage of pests eaten or parasitized in Bt and non-Bt cotton are often similar.

Biological Control Agents

Predators

Predators are mainly free-living species that directly consume a large number of prey during their whole lifetime. Given that many major crop pests are insects, many of the predators used in biological control are insectivorous species. Lady beetles, and in particular their larvae which are active between May and July in the northern hemisphere, are voracious predators of aphids, and also consume mites, scale insects and small caterpillars. The spotted lady beetle (Coleomegilla maculata) is also able to feed on the eggs and larvae of the Colorado potato beetle (Leptinotarsa decemlineata).

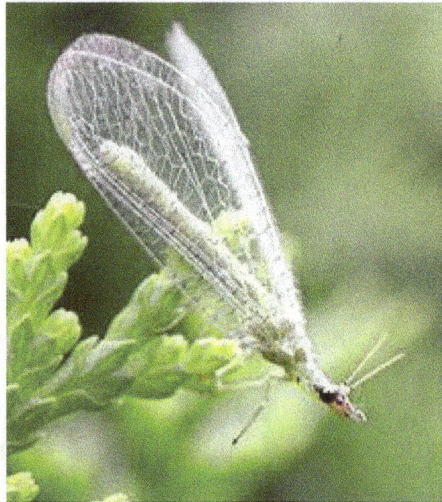

Predatory lacewings are available from biocontrol dealers.

The larvae of many hoverfly species principally feed upon aphids, one larva devouring up to 400 in its lifetime. Their effectiveness in commercial crops has not been studied.

Predatory *Polistes* wasp searching for bollworms or
other caterpillars on a cotton plant.

Several species of entomopathogenic nematode are important predators of insect and other invertebrate pests. Phasmarhabditis hermaphrodita is a microscopic nematode that kills slugs. Its complex life cycle includes a free-living, infective stage in the soil where it becomes associated with a pathogenic bacteria such as Moraxella osloensis. The nematode enters the slug through the posterior mantle region, thereafter feeding and reproducing inside, but it is the bacteria that kill the slug. The nematode is available commercially in Europe and is applied by watering onto moist soil.

Species used to control spider mites include the predatory mites Phytoseiulus persimilis, Neoseilus californicus, and Amblyseius cucumeris, the predatory midge Feltiella acarisuga, and a ladybird Stethorus punctillum. The bug Orius insidiosus has been successfully used against the two-spotted spider mite and the western flower thrips (Frankliniella occidentalis).

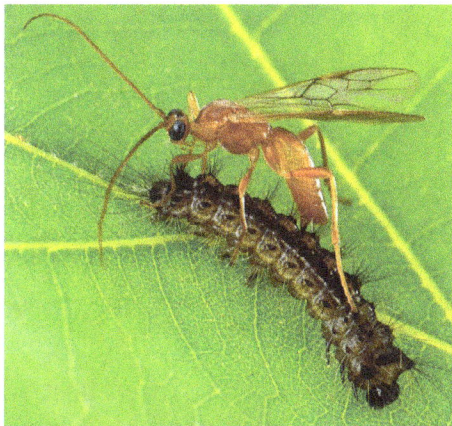

The parasitoid wasp *Aleiodes indiscretus* parasitizing
a gypsy moth caterpillar, a serious pest of forestry.

Predators including Cactoblastis cactorum (mentioned above) can also be used to destroy invasive plant species. As another example, the poison hemlock moth (Agonopterix alstroemeriana) can be used to control poison hemlock (Conium maculatum). During its larval stage, the moth strictly consumes its host plant, poison hemlock, and can exist at hundreds of larvae per individual host plant, destroying large swathes of the hemlock.

For rodent pests, cats are effective biological control when used in conjunction with reduction of harborage/hiding locations. While cats are effective at preventing rodent "population explosions", they are not effective for eliminating pre-existing severe infestations. Barn owls are also sometimes used as biological rodent control. Although there are no quantitative studies of the effectiveness of barn owls for this purpose, they are known rodent predators that can be used in addition to or instead of cats; they can be encouraged into an area with nest boxes.

Parasitoids

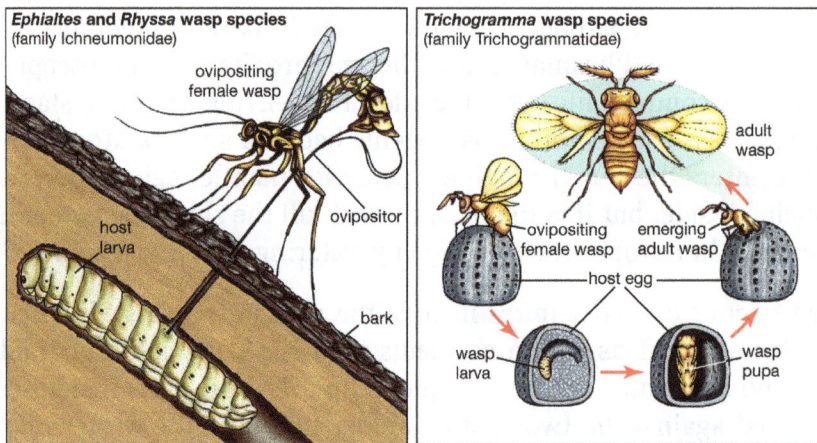

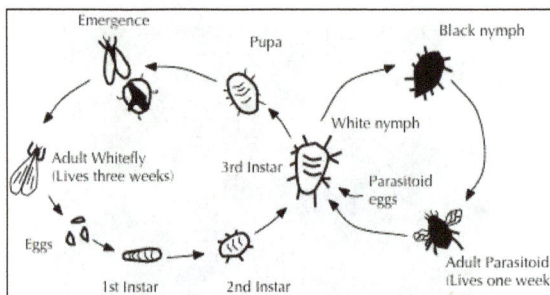

Life cycles of greenhouse whitefly and its parasitoid wasp Encarsia formosa.

Parasitoids lay their eggs on or in the body of an insect host, which is then used as a food for developing larvae. The host is ultimately killed. Most insect parasitoids are wasps or flies, and many have a very narrow host range. The most important groups are the ichneumonid wasps, which mainly use caterpillars as hosts; braconid wasps, which attack caterpillars and a wide range of other insects including aphids; chalcid wasps, which parasitize eggs and larvae of many insect species; and tachinid flies, which parasitize a wide range of insects including caterpillars, beetle adults and larvae, and true bugs. Parasitoids are most effective at reducing pest populations when their host organisms have limited refuges to hide from them.

Parasitoids are among the most widely used biological control agents. Commercially, there are two types of rearing systems: short-term daily output with high production of parasitoids per day, and long-term, low daily output systems. In most instances, production will need to be matched with the appropriate release dates when susceptible host species at a suitable phase of development will be available. Larger production facilities produce on a yearlong basis, whereas some facilities produce only seasonally. Rearing facilities are usually a significant distance from where the agents are to be used in the field, and transporting the parasitoids from the point of production to the point of use can pose problems. Shipping conditions can be too hot, and even vibrations from planes or trucks can adversely affect parasitoids.

Encarsia formosa is a small predatory chalcid wasp which is a parasitoid of whitefly, a sap-feeding insect which can cause wilting and black sooty moulds in glasshouse vegetable and ornamental crops. It is most effective when dealing with low level infestations, giving protection over a long period of time. The wasp lays its eggs in young whitefly 'scales', turning them black as the parasite larvae pupate. Gonatocerus ashmeadi (Hymenoptera: Mymaridae) has been introduced to control the glassy-winged sharpshooter Homalodisca vitripennis (Hemiptera: Cicadellidae) in French Polynesia and has successfully controlled ~95% of the pest density.

The eastern spruce budworm is an example of a destructive insect in fir and spruce forests. Birds are a natural form of biological control, but the *Trichogramma minutum*, a species of parasitic wasp, has been investigated as an alternative to more controversial chemical controls.

There are a number of recent studies pursuing sustainable methods for controlling urban cockroaches using parasitic wasps. Since most cockroaches remain in the sewer system and sheltered areas which are inaccessible to insecticides, employing active-hunter wasps is a strategy to try and reduce their populations.

Pathogens

Pathogenic micro-organisms include bacteria, fungi, and viruses. They kill or debilitate their host and are relatively host-specific. Various microbial insect diseases occur naturally, but may also be used as biological pesticides. When naturally occurring, these outbreaks are density-dependent in that they generally only occur as insect populations become denser.

Bacteria

Bacteria used for biological control infect insects via their digestive tracts, so they offer only limited options for controlling insects with sucking mouth parts such as aphids and scale insects. Bacillus thuringiensis, a soil-dwelling bacterium, is the most widely applied species of bacteria used for biological control, with at least four sub-species used against Lepidopteran (moth, butterfly), Coleopteran (beetle) and Dipteran (true fly) insect pests. The bacterium is available to organic farmers in sachets of dried spores which are mixed with water and sprayed onto vulnerable plants such as brassicas and fruit trees. Genes from B. thuringiensis have also been incorporated into transgenic crops, making the plants express some of the bacterium's toxins, which are proteins. These confer resistance to insect pests and thus reduce the necessity for pesticide use. If pests develop resistance to the toxins in these crops, B. thuringiensis will become useless in organic farming also. The bacterium Paenibacillus popilliae which causes milky spore disease has been found useful in the control of Japanese beetle, killing the larvae. It is very specific to its host species and is harmless to vertebrates and other invertebrates.

Fungi

Entomopathogenic fungi, which cause disease in insects, include at least 14 species that attack aphids. Beauveria bassiana is mass-produced and used to manage a wide variety of insect pests including whiteflies, thrips, aphids and weevils. Lecanicillium spp. are deployed against white flies, thrips and aphids. Metarhizium spp. are used against pests including beetles, locusts and other grasshoppers, Hemiptera, and spider mites. Paecilomyces fumosoroseus is effective against white flies, thrips and aphids; Purpureocillium lilacinus is used against root-knot nematodes, and 89 Trichoderma species against certain plant pathogens. Trichoderma viride has been used against Dutch elm disease, and has shown some effect in suppressing silver leaf, a disease of stone fruits caused by the pathogenic fungus Chondrostereum purpureum.

Green peach aphid, a pest in its own right and a vector of plant viruses, killed by the fungus Pandora neoaphidis (Zygomycota: Entomophthorales) Scale bar = 0.3 mm.

The fungi Cordyceps and Metacordyceps are deployed against a wide spectrum of arthropods. Entomophaga is effective against pests such as the green peach aphid.

Several members of Chytridiomycota and Blastocladiomycota have been explored as agents of biological control. From Chytridiomycota, Synchytrium solstitiale is being considered as a control agent of the yellow star thistle (Centaurea solstitialis) in the United States.

Viruses

Baculoviruses are specific to individual insect host species and have been shown to be useful in biological pest control. For example, the Lymantria dispar multicapsid nuclear polyhedrosis virus has been used to spray large areas of forest in North America where larvae of the gypsy moth are causing serious defoliation. The moth larvae are killed by the virus they have eaten and die, the disintegrating cadavers leaving virus particles on the foliage to infect other larvae.

A mammalian virus, the rabbit haemorrhagic disease virus was introduced to Australia to attempt to control the European rabbit populations there. It escaped from quarantine and spread across the country, killing large numbers of rabbits. Very young animals survived, passing immunity to their offspring in due course and eventually producing a virus-resistant population. Introduction into New Zealand in the 1990s was similarly successful at first, but a decade later, immunity had developed and populations had returned to pre-RHD levels.

Oomycota

Lagenidium giganteum is a water-borne mould that parasitizes the larval stage of mosquitoes. When applied to water, the motile spores avoid unsuitable host species and search out suitable mosquito larval hosts. This mould has the advantages of a dormant phase, resistant to desiccation, with slow-release characteristics over several years. Unfortunately, it is susceptible to many chemicals used in mosquito abatement programmes.

Competitors

The legume vine Mucuna pruriens is used in the countries of Benin and Vietnam as a biological control for problematic Imperata cylindrica grass: the vine is extremely vigorous and suppresses neighbouring plants by out-competing them for space and light. Mucuna pruriens is said not to be invasive outside its cultivated area. Desmodium uncinatum can be used in push-pull farming to stop the parasitic plant, witchweed (Striga).

The Australian bush fly, Musca vetustissima, is a major nuisance pest in Australia, but native decomposers found in Australia are not adapted to feeding on cow dung, which is where bush flies breed. Therefore, the Australian Dung Beetle Project (1965–1985), led by George Bornemissza of the Commonwealth Scientific and Industrial Research Organisation, released forty-nine species of dung beetle, to reduce the amount of dung and therefore also the potential breeding sites of the fly.

Combined use of Parasitoids and Pathogens

In cases of massive and severe infection of invasive pests, techniques of pest control are often used in combination. An example is the emerald ash borer, Agrilus planipennis, an invasive beetle from China, which has destroyed tens of millions of ash trees in its introduced range in North America. As part of the campaign against it, from 2003 American scientists and the Chinese Academy of Forestry searched for its natural enemies in the wild, leading to the discovery of several parasitoid wasps, namely Tetrastichus planipennisi, a gregarious larval endoparasitoid, Oobius agrili, a solitary, parthenogenic egg parasitoid, and Spathius agrili, a gregarious larval ectoparasitoid. These have been introduced and released into the United States of America as a possible biological control of the emerald ash borer. Initial results for Tetrastichus planipennisi have shown promise, and it is now being released along with Beauveria bassiana, a fungal pathogen with known insecticidal properties.

Difficulties

Many of the most important pests are exotic, invasive species that severely impact agriculture, horticulture, forestry and urban environments. They tend to arrive without their co-evolved parasites, pathogens and predators, and by escaping from these, populations may soar. Importing the natural enemies of these pests may seem a logical move but this may have unintended consequences; regulations may be ineffective and there may be unanticipated effects on biodiversity, and the adoption of the techniques may prove challenging because of a lack of knowledge among farmers and growers.

Side Effects

Biological control can affect biodiversity through predation, parasitism, pathogenicity, competition, or other attacks on non-target species. An introduced control does not always target only the intended pest species; it can also target native species. In Hawaii during the 1940s parasitic wasps were introduced to control a lepidopteran pest and the wasps are still found there today. This may have a negative impact on the native ecosystem; however, host range and impacts need to be studied before declaring their impact on the environment.

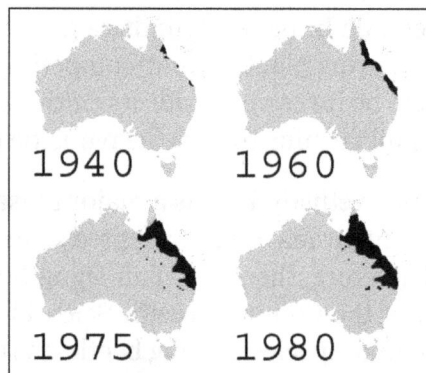

Cane toad (introduced into Australia 1935) spread from 1940 to 1980: it was ineffective as a control agent. Its distribution has continued to widen since 1980.

Vertebrate animals tend to be generalist feeders, and seldom make good biological control agents; many of the classic cases of "biocontrol gone awry" involve vertebrates. For example, the cane toad (Rhinella marina) was intentionally introduced to Australia to control the greyback cane beetle (Dermolepida albohirtum),and other pests of sugar cane. 102 toads were obtained from Hawaii and bred in captivity to increase their numbers until they were released into the sugar cane fields of the tropic north in 1935. It was later discovered that the toads could not jump very high and so were unable to eat the cane beetles which stayed on the upper stalks of the cane plants. However, the toad thrived by feeding on other insects and soon spread very rapidly; it took over native amphibian habitat and brought foreign disease to native toads and frogs, dramatically reducing their populations. Also, when it is threatened or handled, the cane toad releases poison from parotoid glands on its shoulders; native Australian species such as goannas, tiger snakes, dingos and northern quolls that attempted to eat the toad were harmed or killed. However, there has been some recent evidence that native predators are adapting, both physiologically and through changing their behaviour, so in the long run, their populations may recover.

Rhinocyllus conicus, a seed-feeding weevil, was introduced to North America to control exotic musk thistle (Carduus nutans) and Canadian thistle (Cirsium arvense). However, the weevil also attacks native thistles, harming such species as the endemic Platte thistle (Cirsium neomexicanum) by selecting larger plants (which reduced the gene pool), reducing seed production and ultimately threatening the species' survival. Similarly, the weevil Larinus planus was also used to try to control the Canadian thistle, but it damaged other thistles as well. This included one species classified as threatened.

The small Asian mongoose (Herpestus javanicus) was introduced to Hawaii in order to control the rat population. However, the mongoose was diurnal, and the rats emerged at night; the mongoose therefore preyed on the endemic birds of Hawaii, especially their eggs, more often than it ate the rats, and now both rats and mongooses threaten the birds. This introduction was undertaken without understanding the consequences of such an action. No regulations existed at the time, and more careful evaluation should prevent such releases now.

The sturdy and prolific eastern mosquitofish (Gambusia holbrooki) is a native of the southeastern United States and was introduced around the world in the 1930s and '40s to feed on mosquito larvae and thus combat malaria. However, it has thrived at the expense of local species, causing a decline of endemic fish and frogs through competition for food resources, as well as through eating their eggs and larvae. In Australia, control of the mosquitofish is the subject of discussion; in 1989 researchers A. H. Arthington and L. L. Lloyd stated that "biological population control is well beyond present capabilities".

Grower Education

A potential obstacle to the adoption of biological pest control measures is that growers may prefer to stay with the familiar use of pesticides. However, pesticides have unde-sired effects, including the development of resistance among pests, and the destruction of natural enemies; these may in turn enable outbreaks of pests of other species than the ones originally targeted, and on crops at a distance from those treated with pesti-cides. One method of increasing grower adoption of biocontrol methods involves let-ting them learn by doing, for example showing them simple field experiments, enabling them to observe the live predation of pests, or demonstrations of parasitised pests. In the Philippines, early season sprays against leaf folder caterpillars were common prac-tice, but growers were asked to follow a 'rule of thumb' of not spraying against leaf fold-ers for the first 30 days after transplanting; participation in this resulted in a reduction of insecticide use by 1/3 and a change in grower perception of insecticide use.

Sterile Insect Technique

The screw-worm fly was the first pest successfully eliminated from an area through the sterile insect technique, by the use of an integrated area-wide approach.

The sterile insect technique (SIT) is a method of biological insect control, whereby overwhelming numbers of sterile insects are released into the wild. The released in-sects are preferably male, as this is more cost-effective and the females may in some situations cause damage by laying eggs in the crop, or, in the case of mosquitoes, tak-ing blood from humans. The sterile males compete with wild males to mate with the females. Females that mate with a sterile male produce no offspring, thus reducing the next generation's population. Sterile insects are not self-replicating and, therefore, cannot become established in the environment. Repeated release of sterile males over low population densities can further reduce and in cases of isolation eliminate pest populations, although cost-effective control with dense target populations is subjected to population suppression prior to the release of the sterile males.

The technique has successfully been used to eradicate the screw-worm fly (Cochliomyia hominivorax) from North and Central America. Many successes have been achieved for control of fruit fly pests, most particularly the Mediterranean fruit fly (Ceratitis capita-ta) and the Mexican fruit fly (Anastrepha ludens).

Sterilization is induced through the effects of irradiation on the reproductive cells of the insects. SIT does not involve the release of insects modified through transgenic (genetic engineering) processes. Moreover, SIT does not introduce non-native species into an ecosystem.

Drawbacks

- Naturally low population periods or repeated pesticide treatment are sometimes required to suppress populations before the use of sterile insects.

- Sex separation can be difficult, though this can be easily performed on a large scale where genetic sexing systems have been developed as for the Mediterranean fruit fly.

- Radiation, transport and release treatments can reduce male mating fitness.

- The technique is species-specific. For instance, the technique must be implemented separately for each of the 6 economically important tsetse fly species.

- Mass rearing and irradiation require precision processes. Failures have occurred when unexpectedly fertile breeding males were released.

- Area-wide approach is more effective, as migration of wild insects from outside the control area could recreate the problem.

- The cost of producing sufficient sterile insects can be prohibitive in some locations but decreases with economies of scale.

Economic Benefits

Economic benefits have been demonstrated. The direct benefits of screwworm eradication to the North and Central American livestock industries are estimated to be over $1.5 billion/year, compared with an investment over half a century around $1 billion. Mexico protects a fruit and vegetable export market of over $3 billion/year through an annual investment around $25 million. Medfly-free status has been estimated to have opened markets for Chile's fruit exports up to $500 million. Eradication of tsetse has resulted in major socioeconomic benefits for Zanzibar. When implemented on an area-wide basis and a scaled rearing process, SIT is cost-competitive with conventional control, in addition to its environmental benefits.

Mechanical Pest Control

The reduction or suppression of insect populations by means of manual devices is referred to as mechanical control.

The mechanical control involves the use of the following tactics:

- Handpicking: Handpicking and destruction of large sized, conspicuous, immature or mature stages of insects is the most ancient method. It can prove fairly effective under certain conditions. This is still being used for picking out lice from human hair (at home). In the field, insects can be handpicked if they are:

  ○ Easily accessible to the picker,

  ○ Large and conspicuous,

  ○ Present in large numbers.

Collection and destruction of egg masses of top borer of first and second brood during March and May in ratoon (sprouted from roots) and autumn-planted sugarcane crops, reduces top borer damage in endemic pockets. Destruction of infested cane stocks harboring larvae of borers is useful for reducing the incidence of these borers.

Handpicking is also generally useful for the management of hairy caterpillars, leaf rollers, tobacco caterpillar, cabbage butterfly, mustard sawfly, Epilachnn beetle, white grubs etc.

Collection and destruction of fallen infested fruits is
effective against fruit flies and fruit barriers.

Manual removal of pink boll worm attacked flowers, withered and drooped terminals infested by spotted boll worms, shed floral bodies attacked by bollworms, leaves with egg masses or younger instars of tobacco caterpillars can reduce the incidence of these pests in cotton crop.

Exclusion by Screens and Barriers

Mechanical exclusion: Mechanical, exclusion consists of the use of devices by which insects are physically prevented from reaching crops and agricultural produce. The various methods include:

- The application of a fluffy cotton band 6" wide, or a band of a sticky material or a band of slippery sheets like alkathene around the tree trunk of a mango tree to prevent the upward movement of the mango mealy bug.

- Screening windows, doors and ventilators of the house to keep away houseflies and mosquitoes, bugs etc. In the morning and at dusk when mosquitoes gather on the screen they can be squashed with a piece of cloth. Screening placed over cavities and windows serves to exclude termites.

- Packaging is a barrier to prevent insect infestation. Polymer films, and laminations can protect packages from insect infestation, whereas polyester, polyurethane resist insect penetration. Wrapping individual fruits of pomegranate and citrus with butter paper envelope to save them from attack of the anaar butterfly and fruit-sucking moths respectively. Fruits wrapped in paper bags or cloth or straw remain protected to the extent of nearly 95 per cent against fruit flies. Maize cobs can be protected from the attack of crows if the nearest leaf is wrapped around thedexposed portion of the cob.

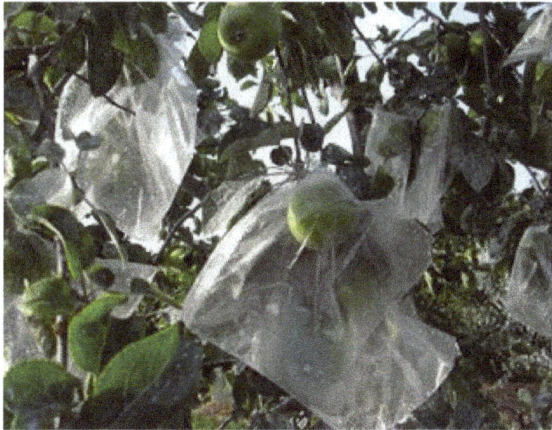

Guava fruits protected by polythene bags against fruit flies and other pests.

- Digging trenches: Digging of 30 to 60 cm deep trenches or erecting 30 cm high tin sheet barriers around field is useful for protecting them from moving bands of locusts and hairy caterpillars.

- Use of ant pans to prevent ants from ascending.

- Lighting: Using red light in the monsoons to keep away most of insects, and to keep the field well lit with white light at night to protect it against certain insects.

- Light reflection by aluminum foil is effective against aphids. Similarly, light reflected by plastic ribbon bands or plastic flags hung in the ripening rice fields will protect the crop from bird attack.

- Scaring birds by creating noise with explosives; an automatic device is available in which an explosive gas catches fire intermittently and a loud noise is produced. Model figures of predators and raptors can be placed in areas of high visibility to discourage vertebrate pests or farmers themselves can scare birds ways from.

- Row covers used in broccoli and cabbage production reduces damage by its pests. Floating row cover (FRC) alongwith transparent polyethylene mulch (TPM) are found to be effective against whiteflies and aphids.

- Trenches: Plastic lined trenches were found to restrict the movement of beetles in potato fields.

- Particle barriers: Materials such as sand, granite, glass splinters can act as termite barriers.

- Inert dusts are used to protect the grain from stored grain insects. These dusts are non-silica dusts, sand or silica aero gels. Though these dusts act slowly, however their mammalian toxicity is low and develop negligible resistance.

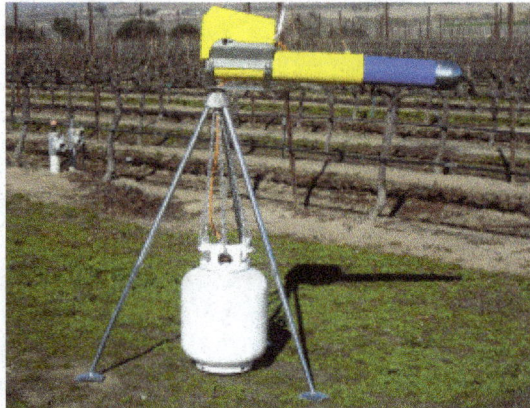

Explosive propane-fueled sound cannon as a sound based scare tactic for birds.

A farmer scaring birds away from maize crop.

Trapping and Suction Devices

Several types of mechanical devices are used for collecting insect pests. Insects are attracted by suitable baits into cages, from which their exit is difficult. Traps have been in use since long as effective control measure for insects. In the past, traps the sole method of controlling pests. Other than insect control, traps also provide of the pest to lower the valuable information for:

- Estimating pest intensity/crop loss assessment.

- Monitoring initial infestation.

- Monitoring the periodicity of pest activities.

Various types of traps have been devised for collecting and killing different types of insects:

- The cricket trap: A deep cylindrical vessel containing beer as a bait and having wooden splinters to aid crickets to reach the bottom.

- House-fly trap: A box, containing a piece of stale cake, with a side opening for the insects to get in only to be trapped in a wire gauze cage on the top.

- Light traps: Light traps for attracting and mass killing of several species of moths and beetles were used as a control measure before the advent of synthetic organic insecticides. The traps could still be useful for monitoring the population of important pests in an area. Trapping of adults through light traps has proved encouraging in controlling top borer, root borer and white grub damage in sugarcane, red hairy caterpillar and ber beetle. A typical light trap uses an ultraviolet fluorescent tube above a collecting jar containing a killing agent UV light has the wavelength from 320-380 nm. Light trap efficiency is improved by placing it in dark areas.

- Air suction traps: These traps are fixed in godowns against stored grain pests. Suction light traps are very effective for those insects which form a group or some supporting surface like windows e.g. Cigarette beetles which congregate together, during sunsel or cloudy days can be trapped by this method.

- Electrocuting trap: Live metal screens on which insects are electrocuted. These traps have been efficiently used to reduce housefly populations as you must have seen their placement in restaurants.

- Plant materials as traps: Compacted plant materials such as sprouts or milkweed placed in fields are used to control cut worms.

- Fermentation traps: Pheromone baits used in traps are being used for reducing lepidopteran (moths and butterfly) pests. Moths are naturally attracted to molasses, fermenting fruit, tree sap, honeydew and flower nectar.

- Colured traps: Different insects respond to different colours. The selection of the colours depend upon the position of traps, physiological stage of insect and quality of the incident wavelengths hitting the traps.

- Trap versus vertebrates: Trapping has been used to control vertebrates pests for thousands of years. Their are two main types of traps for vertebrates. One type kills the animals, such as the Macabee trap for killing gophers. The trap is placed in the burrow and catches the gopher as it crawls through the trap. One problem with kill traps is that they do not always kill the animal the other type is live trap, because they catch the animal without killing it. The household mouse trap is familiar example of this type.

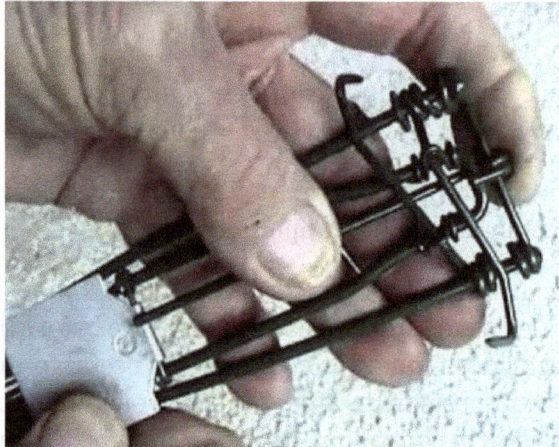

Setting a Macabee trap.

Rat caught in a live trap.

Table: Response of insects to different trap colours.

Insect	Trap colours
Aphid	Yellow
Thrips	White

Mexican fruit fly	Green, Yellow
Apple maggots	Red, Yellow
Leafhoppers	Orange
Onion fly	White, Blue

Banding of fruit trees: This is effective against caterpillars ants and mealy bugs which crawl up the trunk from the soil to the shoots. The banding material like grease smeared around trunk trees arrests the crawling insects below the banded part where either they die of starvalion or by getting entangled with the sticky banding material.

Use of Hand-nets and Bag-nets

The collection of adults with hand nets is recommended for Pyrilla (pest of sugar cane), when these insects are migrating in April- May from maize to sugarcane.

The field bag is a strong cloth bag, 2 meters long with its mouth measuring 1 × 1.5 meters supported with bamboo sticks and two strings on the upper side. It is scraped on the surface of the ground by two men and is recommended against surface grasshoppers, rice grasshoppers, crickets etc. Even a one-man bag can be devised by reducing the size of its mouth.

Clipping, Pruning and Crushing

Pruning and destruction of infested shoots and floral parts is effective in checking the multiplication of scales, mealy bugs and gall midges attacking fruit trees like grapes, citrus, ber, fig, custard apple etc.

A useful approach for the management of mustard aphid infesting Brassicu (mustard) crop involves clipping and destruction of aphid-infested twigs.

Beating and Hooking

Killing houseflies with fly wrappers and locusts with brooms or thorny bushes is effective. On coconut palms, the borer can be picked out of the holes with the help of crooked hooks made of iron.

Shaking or Jarring

Shaking small trees and shrubs, particularly early in the morning in the cold season when the insects are benumbed, and collecting them in open tubs containing kerosinized water or simpfy burying them in pits is effective against locust and the defoliating beetles.

Hopperdozer

Hopperdozer is a physical technique used to control grasshoppers by catching them. Based on the leaping behaviour of grasshoppers, hopperdozer is designed. Hopperdozer has been in use since early nineties.

Sieving and Winnowing

These are commonly employed against insect pests of stored grains. A good number are removed with these operations, particularly the grubs of Tribolium and trogoderma, which infest wheat.

Flaming and Burning

Flaming and burning are methods which use heat to kill insects and to eliminate breeding sites. Flaming, with kerosene as the fuel, started in US to control chinch bugs and the green bug. The burning of locust adults and hoppers with the help of flamethrowers and flame torches, although costly, has a good psychological effect in mobilizing the public for locust control operations. In Suriname agricultural land was developed from rain forests by burning and flaming.

Burning of rain forests in Suriname

Physical Pest Control

Physical Pest Control is a method of getting rid of insects and small rodents by killing, removing, or setting up barriers that will prevent further destruction of one's plants.

These methods are used primarily for crop growing, but some methods can be applied to homes as well.

Methods

Barriers

Row covers are useful for keeping insects out of one's plants, typically used for horticultural crops. They are made out of either plastic or polyester. They are made thin and light to allow plants to still absorb sunshine and water from the air.

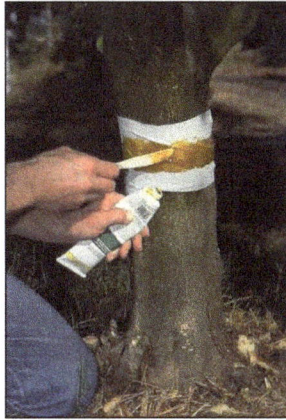

Dog control van.

Diatomaceous earth, made from fossilized and pulverized silica shells, can be used in order to damage the protective cuticle layer of insects that have them, such as ants. When this layer is damaged, the insects become vulnerable to drying out. Unfortunately, the effectiveness of Diatomaceous earth decreases if it is wet. Therefore, it must be used often. This method was used back in the 1930s and 1940s when farmers would run dust over their fields. This would have the very same effect as diatomaceous earth.

Fire

For farmers, fire has been a powerful technique used to destroy insect breeding grounds. It is used to burn the top of the soil in order to kill the insects that lie there. Unfortunately, this can present some drawbacks. Fire can make the soil much less effective or get rid of the insects that are beneficial to the plants. Also, there is no guarantee that it will actually solve the pest problems since there may be larvae below the surface of the soil.

Firearms

Historically, firearms have been one of the primary methods used for pest control. "Garden Guns" are smooth bore shotguns specifically made to fire .22 caliber snake shot or 9mm Flobert, and are commonly used by gardeners and farmers for pest control. Garden Guns are short range weapons that can do little harm past 15 to 20 yards, and they're relatively quiet when fired with snake shot, compared to a standard ammunition. These guns are especially effective inside of barns and sheds, as the snake shot will not shoot holes in the roof or walls, or more importantly injure livestock with a ricochet. They are also used for pest control at airports, warehouses, stockyards, etc.

The most common shot cartridge is .22 Long Rifle loaded with #12 shot. At a distance of about 10 feet (3 m), which is about the maximum effective range, the pattern is about 8 inches (20 cm) in diameter from a standard rifle. Special smoothbore shotguns, such as the Marlin Model 25MG can produce effective patterns out to 15 or 20 yards using .22 WMR shotshells, which hold 1/8 oz. of #12 shot contained in a plastic capsule.

Animals

Dogs, cats, ferrets, mongoose and other animals have been historically used for pest control.

The Rat Terrier is an American dog breed with a background as a farm dog and hunting companion. Specifically bred for killing rats, today's Rat Terrier is an intelligent and active small dog that is kept both for pest control and as a family pet. Cats are also valued for companionship and for their ability to hunt vermin. Ferrets are used for hunting, or ferreting. With their long, lean build, and inquisitive nature, ferrets are very well equipped for getting down holes and chasing rodents, rabbits and moles out of their burrows. Mongooses have long been celebrated for their ability to handle venomous snakes, as immortalized in the short story *Rikki-Tikki-Tavi*.

Temperature Control

Placing produce inside of cold storage containers lengthens how long the produce lasts while also hindering the growth of insects inside of them. Another method to use is to heat, as it will kill the insect larvae in certain types of produce. An example would be with mangoes, where they are placed into a hot water bath in order to kill any eggs and larvae.

Traps

Fly paper or sticky boards are devices used in order to capture insects as they land upon the surface of the trap. They are covered in a substance that attracts insects, but are actually very sticky or poisonous. These traps are commonly used for flies or leafhoppers.

Trap strips are crops that are grown on fields with the intention of using them to attract insects and not have insects infest the other crops that are being grown. The insects can then be dealt with much more easily than if they were to have been spread throughout an entire field. Trap strips are very useful for dealing with the wheat stem sawfly. The sawflies will go only as far as they need to in order to plant their eggs. If solid stemmed plants are planted around the a crop field, then that's where the sawflies will go and the sawflies' larvae can't survive in the solid stem.

Large Scale Usage

On a much larger scale, physical control methods become much less effective because of the time that must be invested into it and because it is likely to be less economical. For example, taking care of a single tree is simple, but taking care of 500, like on a farm, would be impossible using physical control.

Fumigation

Fumigation is a method of pest control that completely fills an area with gaseous pesticides—or fumigants—to suffocate or poison the pests within. It is used to control pests in buildings (structural fumigation), soil, grain, and produce, and is also used during processing of goods to be imported or exported to prevent transfer of exotic organisms. This method also affects the structure itself, affecting pests that inhabit the physical structure, such as woodborers and drywood termites.

Process

Fumigation generally involves the following phases: First the area intended to be fumigated is usually covered to create a sealed environment; next the fumigant is released into the space to be fumigated; then, the space is held for a set period while the fumigant gas percolates through the space and acts on and kills any infestation in the product, next the space is ventilated so that the poisonous gases are allowed to escape from the space, and render it safe for humans to enter. If successful, the fumigated area is now safe and pest free.

Tent Fumigation

Structural fumigation techniques differ from building to building, but in houses a rubber (usually the tent is made of plastic\pvc coated canvas material) tent(s) are often placed over the entire house while the pesticides are being released into the vacant residence. This process is called tent fumigation or "tenting". The sealed tent concentrates the poisonous gases and prevents them from escaping into the neighborhood. The process can take up to a week depending on the fumigant used, which is in turn dependent on the temperature and the pest.

Operating Theatre

Fumigation of hospital rooms with high concentrations of toxic chemicals has been proposed to reduce microbial agents on hospital surfaces and to control surgical site infections. For this formaldehyde fumigation has long been an accepted method for areas where microbiological cleanliness is required. Fumigation with formaldehyde vapor is the recognized and most commonly used method because of its cost effective procedure. However, due to safety and efficacy concerns there is a need for alternative methods. Vaporized Hydrogen Peroxide is a dry gaseous method that has been used as a reliable alternative for aseptic processing isolators and more recently for room/ facility decontamination.

Chemicals

- Methyl bromide was among the most widely used fumigants until its production and use was restricted by the montreal protocol due to its role in ozone depletion,

- 1,3-Dichloropropene,

- Dazomet (methyl isothiocyanate precursor),

- Chloropicrin,

- DBCP,

- Formaldehyde,

- Hydrogen cyanide,

- Iodoform,

- Methyl isocyanate,

- Phosphine,

- Sulfuryl fluoride.

Safety

Fumigation is a hazardous operation. Generally it is a legal requirement that the operator who carries out the fumigation operation holds official certification to perform the fumigation as the chemicals used are toxic to most forms of life, including humans.

Post operation ventilation of the area is a critical safety aspect of fumigation. It is important to distinguish between the pack or source of the fumigant gas and the environment which has been fumigated. While the fumigant pack may be safe and spent, the space will still hold the fumigant gas until it has been ventilated.

- Post Fumigation: This stage is very important because people need to know how long after fumigation is safe. After dismantling fumigation tents and aerating the house, a clearance device is used to test the level of dissipation of the fumigant. Clearance devices are electronic devices which are calibrated to measure toxicity dissipation levels of fumigants. Your home is only safe after being certified by pest control technicians as safe.

References

- Pest-management: napawatersheds.org, Retrieved 1 January, 2019

- Georghiou, g.p. (2012). Pest resistance to pesticides. Springer science & business media. Pp. 1–3. Isbn 978-1-4684-4466-7

- "Pesticides". National institute of health sciences. National institute of environmental health. Retrieved 5 april 2013

- Lieutier, françois; day, keith r.; battisti, andrea; grégoire, jean-claude; evans, hugh f. (2007). Bark and wood boring insects in living trees in europe, a synthesis. Springer. P. 39. Isbn 978-1-4020-2241-8

- "Biological control: paenibacillus popilliae". Cornell university. Archived from the original on 21 june 2016. Retrieved 15 june 2016

- Kuris, armand m. (march 2003). "did biological control cause extinction of the coconut moth, levuana iridescens, in fiji?". Biological invasions. 5 (1): 133–141. Doi:10.1023/a:1024015327707

- Peter, k. V. (2009). Basics of horticulture. New india publishing. P. 288. Isbn 978-81-89422-55-4. Archived from the original on 2017-04-07

- Types-of-pests : nios.ac.in, Retrieved 19 July, 2019

- Dyck, v.a.; hendrichs, j.; robinson, a.s., eds. (2005). Sterile insect technique: principles and practice in area-wide integrated pest management. Dordrecht, the netherlands: springer. Isbn 1-4020-4050-4

- Chen, lin h.; hamer, davidson h. (2016). "zika virus: rapid spread in the western hemisphere". Annals of internal medicine. 164 (9): 613–5. Doi:10.7326/m16-0150. Issn 0003-4819. Pmid 26832396

- Eger, christopher (28 july 2013). "marlin 25mg garden gun". Marlin firearms forum. Outdoor hub llc. Retrieved 17 september 2016

- Byrns, g.; fuller, t. P. (2011). "the risks and benefits of chemical fumigation in the health care environment". Journal of occupational and environmental hygiene. 8 (2): 104–12. Doi:10.1080/15 459624.2011.547453. Pmid 21253983

Permissions

All chapters in this book are published with permission under the Creative Commons Attribution Share Alike License or equivalent. Every chapter published in this book has been scrutinized by our experts. Their significance has been extensively debated. The topics covered herein carry significant information for a comprehensive understanding. They may even be implemented as practical applications or may be referred to as a beginning point for further studies.

We would like to thank the editorial team for lending their expertise to make the book truly unique. They have played a crucial role in the development of this book. Without their invaluable contributions this book wouldn't have been possible. They have made vital efforts to compile up to date information on the varied aspects of this subject to make this book a valuable addition to the collection of many professionals and students.

This book was conceptualized with the vision of imparting up-to-date and integrated information in this field. To ensure the same, a matchless editorial board was set up. Every individual on the board went through rigorous rounds of assessment to prove their worth. After which they invested a large part of their time researching and compiling the most relevant data for our readers.

The editorial board has been involved in producing this book since its inception. They have spent rigorous hours researching and exploring the diverse topics which have resulted in the successful publishing of this book. They have passed on their knowledge of decades through this book. To expedite this challenging task, the publisher supported the team at every step. A small team of assistant editors was also appointed to further simplify the editing procedure and attain best results for the readers.

Apart from the editorial board, the designing team has also invested a significant amount of their time in understanding the subject and creating the most relevant covers. They scrutinized every image to scout for the most suitable representation of the subject and create an appropriate cover for the book.

The publishing team has been an ardent support to the editorial, designing and production team. Their endless efforts to recruit the best for this project, has resulted in the accomplishment of this book. They are a veteran in the field of academics and their pool of knowledge is as vast as their experience in printing. Their expertise and guidance has proved useful at every step. Their uncompromising quality standards have made this book an exceptional effort. Their encouragement from time to time has been an inspiration for everyone.

The publisher and the editorial board hope that this book will prove to be a valuable piece of knowledge for students, practitioners and scholars across the globe.

Index

www.ingramcontent.com/pod-product-compliance
Lightning Source LLC
Chambersburg PA
CBHW080241230326
41458CB00096B/2771

* 9 7 8 1 6 4 7 4 0 0 6 0 6 *